Portrait of Lieutenant-Colonel
The Hon. Charles Stuart
of the 26th Regiment of Foot
by George Romney, painted
ca. 1779–80.

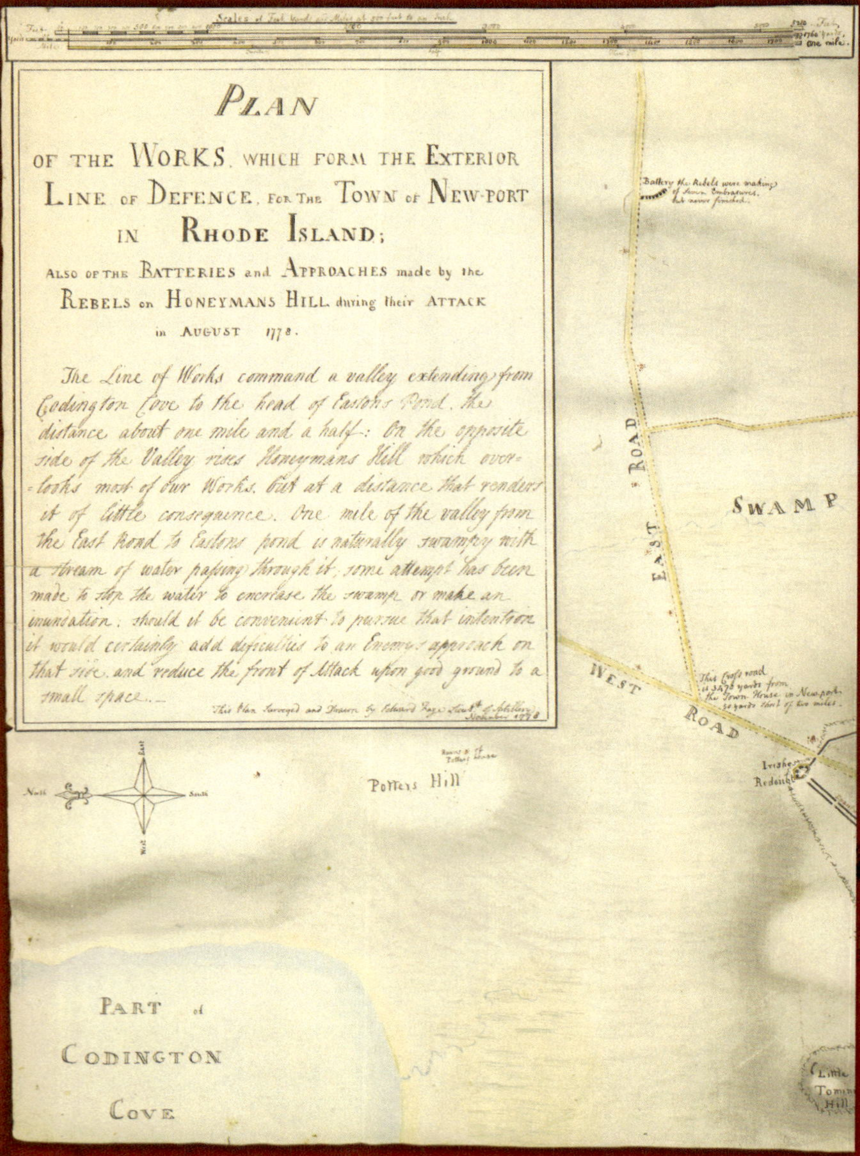

Manuscript map of Newport and colonial approach and defense in August 1778

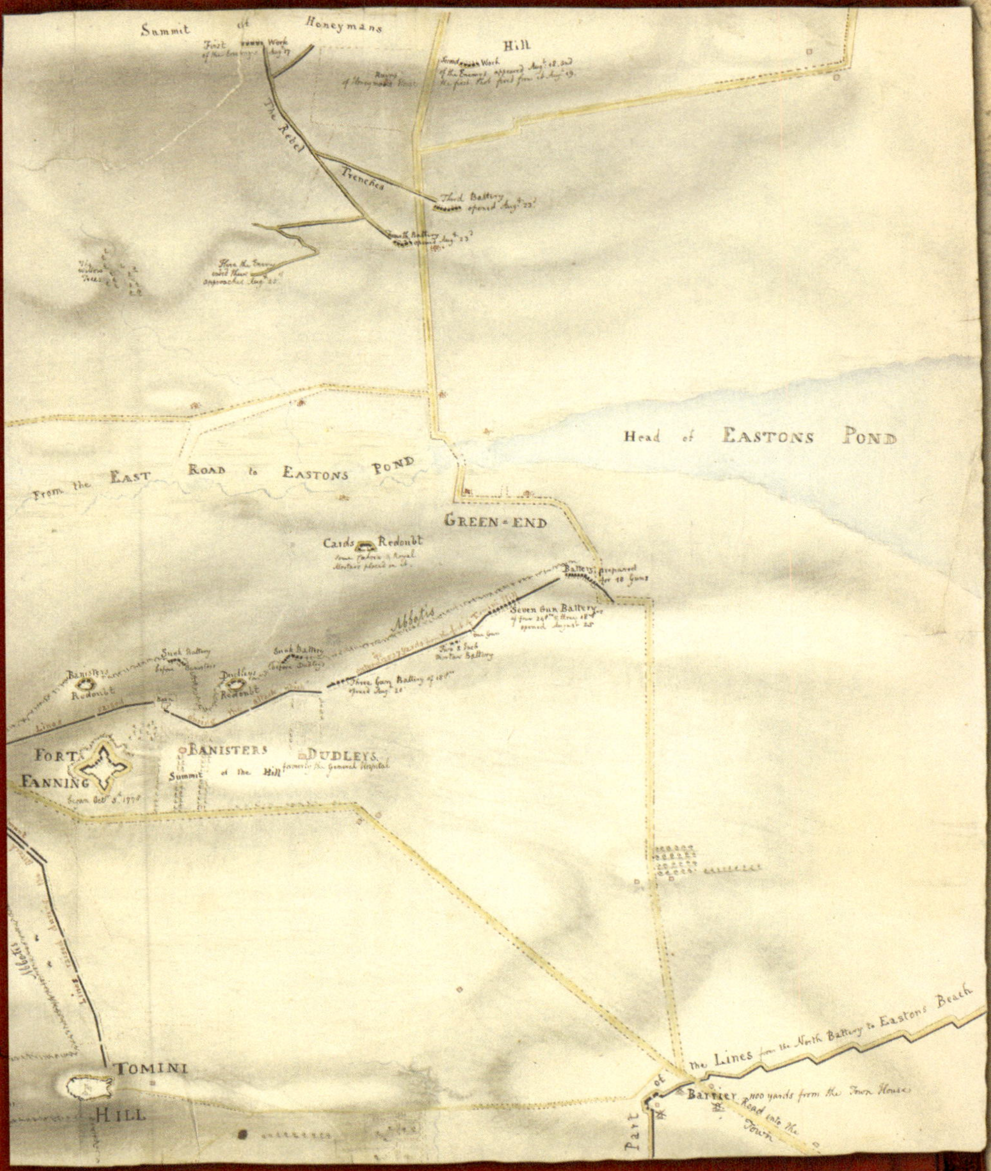

Map surveyed and drawn by
Lieutenant Edward Fage,
Royal Artillery,
November 1778

Manuscript map of Aquidneck Island attributed to Hessian cartographer J. C. Schiffer, ca. 1777

Lieutenant-Colonel The Hon. Charles Stuart's Diary

IN RHODE ISLAND 29 JULY TO 28 AUGUST 1778

A Facsimile,
Transcribed, Annotated, and Edited
with an Introduction and Additional Letters

by JOHN B. HATTENDORF
D.Phil., D.Litt. (Oxon.)
Historian, Rhode Island Society Sons of the Revolution

Rhode Island Society Sons of the Revolution

2024

STONE TOWER PRESS

Published in 2024
By Stone Tower Press
For the Rhode Island Society Sons of the Revolution

Stone Tower Press
7 Ellen Rd.
Middletown, RI 02842
stonetowerpress.com

ISBN: 979-8-9919831-0-5

Cover: Portrait of Sir Charles Stuart by George Ramsey, Glasgow Art Gallery and Museum, Public Domain

Manuscript map of Newport and colonial approach and defense in August 1778, surveyed and drawn by Lieutenant Edward Fage, Royal Artillery, November 1778, William L. Clements Library, University of Michigan (#8373).

Manuscript map of Aquidneck Island attributed to Hessian cartographer J. C. Schiffer, ca. 1777, Library of Congress (#75690704)

Formatting and cover design by Amy Cole, JPL Design Solutions

Printed in the United States of America

Table of Contents

List of Illustrations

Acknowledgments

The editor acknowledges with sincere gratitude the permission granted to reproduce the diary and the photographs in this volume by Stephen Venuti, President, Barrington Preservation Society, and the numerous helpful contributions that Dr. Nathaniel L. Taylor provided in facilitating the photo of copying the manuscript and in drafting and revising parts of the introduction. In addition, we are grateful for the joint work of Dr. Taylor and Maria Bruce in researching and writing a section of the introduction. Dr. John Houlding provided valuable information from his Database of Georgian Officers. Professor Gregory Urwin of Temple University generously responded to the editor's queries and kindly provided additional information from Stephen Gilbert. Terese Austin of the William L. Clements Library made a most welcome and notable contribution by identifying and providing a digital copy of the document in the Clements' collection by Lieutenant Edward Fage.

J.B.H.
Newport, July 2024

Introduction

Charles Stuart was born into a notable family in January 1753 at Kenwood House in Hampstead, London, and eventually rose to be a Lieutenant-General in the British Army before his early death at about the age of 48 in March 1801. Charles was the fourth and favorite son of John Stuart, the third Earl of Bute (1713–1792), Prime Minister of Great Britian in 1762–63, and Mary Wortley Montagu (1718–1794), daughter of Edward Wortley Montagu (1678–1761), Whig politician, and one-time ambassador to the Ottoman Empire, and the writer Lady Mary Wortley Montague, née Pierrepont (1689–1762).

Charles's father, Lord Bute, rose to prominence after being elected in 1737 as one of the Scottish peers to sit in the House of Lords in London, having succeeded at the age of ten to the earldom of Bute on the death of his father in 1723. By 1747, he had become a close friend of Frederick, Prince of Wales. On Frederick's death in 1751, he became tutor to Frederick's eldest son, the new Prince of Wales and future King George III. Due to the close relationship that developed between tutor and student, Lord Bute remained a close advisor when George III came to the throne in 1760 and was widely regarded as the king's favorite. As a Tory, Bute maneuvered himself into power as Secretary of State for the Northern Department in 1761–62 and then Prime Minister in 1762–1763, briefly interrupting the Whig dominance of the ministry between 1714 and 1783. As the political tide turned in Parliament, Bute fell out of favor with the king and resigned as Prime Minister shortly after negotiating the Peace of Paris that ended the Seven Years War in 1763.[1] In his retirement, Lord Bute acquired land on the cliffs of Dorset,

1 Romney Sedgwick, ed., *Letters from George III to Lord Bute, 1756–1766* (London: Macmillan and Co., Ltd, 1939), pp. xliii-lxviii.

with spectacular sea views overlooking the Channel with the Isle of Wight in view to the east. Here, he built a Georgian-style mansion designed by Robert Adam that he called High Cliff House. The grounds contained a smaller version of the Royal Botanical Garden at Kew, which Lord Bute had been instrumental in developing.[2]

Career. The young Charles grew up in a household with several notable siblings who became prominent figures in later life. His eldest brother, John Lord Mountstuart, succeeded to the earldom, became ambassador to Spain, and eventually rose to be the first marquess of Bute. A younger brother, William, became the Church of Ireland's Archbishop of Armagh, and his sister, Lady Luisa, became an author. His brothers—Frederick, James Archibald, and John Lord Mountstuart—all served as Members of Parliament.

Initially educated at home, Charles attended Dr. Graffiani's academy in Kensington for a time. In 1768, he accompanied his father on travels on the continent. Just as he was about to turn 16, he obtained a free death-vacancy commission as an Ensign in the 37th Foot, then commanded by a relative, Lieutenant-General James Stuart, at Minorca on 23 November 1768. Nearly two years later, he was promoted to Lieutenant in the 7th Foot in England, on 13 September 1770, by purchase. Serving under the command of Colonel Richard Prescott, who later served in New England as a Major-General, Stuart's fellow lieutenants included John André. He then purchased his commission as Captain in the 35th Foot in England on 12 March 1773.[3]

---

2 Karl Wolfgang Schweizer, "Stuart, John, third earl of Bute (1713–1792)," *Oxford Dictionary of National Biography* https://www.oxforddnb.com/display/10.1093/ref:odnb/9780198614128.001.0001/odnb-9780198614128-e-26716; Highcliffe Castle website. https://www.highcliffecastle.co.uk/visit/history/

3 Desmond Gregory, "Stuart, Sir Charles (1753–1801)," *Oxford Dictionary of National Biography*. https://doi.org/10.1093/ref:odnb/26697

In 1775, Stuart sailed to North America with the 35th Foot and arrived in Boston on 12 June. The regiment's flank companies fought at Breed's Hill five days later. Serving with the 35th during the siege of Boston for only a few months. Shortly after the battle of Bunker Hill, through the influence of his sister's husband, Colonel Hugh, Earl Percy, he became a grenadier, replacing a company captain, James Lyon, on 30 June 1775. He purchased a commission as major of the 43rd Regiment of Foot, on October 8, 1775. On the British evacuation of Boston in March 1776, he sailed for Halifax, Nova Scotia. On 14 May 1776, General Sir William Howe appointed him second in command of the 2nd Grenadier Battalion, under Lieutenant-Colonel The Hon. Henry Monckton of the 45th Foot. Back in England, while he was in Halifax, he succeeded his elder brother as Member of Parliament for Bossiney, a hamlet of 16 houses in the parish of Tintagel, Cornwall, and he held the post from May 1776 until 1790. After leaving Nova Scotia on 10 June 1776, he took part in the British Army's landings on Staten Island and Long Island in June. In August,[4] he was appointed to command the 4th Grenadier battalion. Undermanned, the 4th Grenadier battalion was disbanded in the middle of the New York campaign. On 8 October 1776, the 4th Grenadier Battalion disbanded, and Stuart returned to being a major in the 43rd Foot.[5]

In November1777, Stuart and his battalion were ordered to Rhode Island under General Sir Henry Clinton. Shortly after Stuart's arrival in Rhode Island, Howe ordered most of the battalion back to New York, leaving Stuart and two companies in Newport. Stuart interpreted this

4 Orderly Books provide varying dates of the appointment—6, 16 or 20 August 1776. Information from Stephen Gilbert provided by Professor Gregory Urwin email to Hattendorf, 1 April 2024.

5 Professor Gregory Urwin email to Hattendorf, 1 April 2024 on Redcoat Images No. 1,245 (Revisited and Addendum); Dr John Houlding Database of Georgian officers; The Hon. Mrs. E. Stuart-Wortley. *A Prime Minister and His Son: From the Correspondence of the 3rd Earl of Bute and of Lt.-General The Hon. Sir Charles Stuart, K.B.* (London: Dutton, 1925), pp. 66–90.

as an attempt by the staff officers of the Commander in Chief, General Sir William Howe, to isolate Stuart as some form of political payback directed at his father, Lord Bute. Seeing it also as a personal, professional disgrace, Stuart wrote to Howe complaining of the "the very cavalier manner in which he had been treated" and requesting leave so that he could return home. to England.[6] General Richard Prescott, then commanding the British troops in Rhode Island, approved of Stuart's biting letters to Howe. One of Stuart's friends, Colonel Allan Maclean, observed "I apprehend they will not wish him to go home. He is a fine young fellow, and I have lost my skill if he will not be one of the best officers in the King's service. He is very clever, exceedingly intelligent, takes great pains, and is as bold as a lion."[7]

This experience served to encourage Stuart's predilection for abrasive criticism of the conduct of the war in America. After temporarily settling down in Rhode Island, Stuart took command of the 3rd Grenadiers on 7 December 1776, when its commanding officer, Major James Marsh of the 57th Foot, was promoted. Stuart remained in Newport until March 1777, when Howe's staff, preparing to march the British Army from New York through New Jersey to Philadelphia, recalled Stuart and the 3rd Grenadiers from Newport. Just before their recall to New York in March 1777, Lord Percy placed Major Stuart of the 43rd Foot in

6 Anonymous. *New Records of the American Revolution: The Letters, Manuscripts and Documents sent by Lieut.-General Sir Charles Stuart, to his Father, the Earl of Bute, 1775–79; also Letters of General Howe, General Clinton, and other Officers to Sir Charles Stuart during the Revolution, 1779–81* (London: Privately Printed by L. Kashnor, 1927), pp. xvi, 21–22.

7 Quoted in Gregory, Desmond. "Stuart, Sir Charles (1753–1801)," *Oxford Dictionary of National Biography*. https://doi.org/10.1093/ref:odnb/26697; Anonymous. *New Records of the American Revolution*, p. 26.

command of the Light Company and Grenadiers of the 54th Foot with the company of Chasseurs.[8]

After Stuart's return to New York from Newport, Howe permanently disbanded the 3rd Grenadiers on 23 March 1777 and merged them with other Grenadier battalions to rationalize manning. Howe did not give Stuart a new command, but he nevertheless participated in the Philadelphia campaign as a supernumerary officer with the 2nd Battalion of Grenadiers at Brandywine. At the center of the action on 11 September 1777, Stuart went into the engagement with the light infantry battalions that formed the vanguard of the Left Division commanded by Major-General Charles Lord Cornwallis.[9] Stuart wrote his father, "A ball gave me a pretty severe scratch in the cheek, and another went through the crown of my hat at the Brandy-Wine, or else I have escaped full well from all the dangers we have been in."[10]

Stuart went as a supernumerary with the 1st Light Infantry Battalion as they fought toward Birmingham and was later involved in negotiations with the rebel defenders prior to attack on Fort Mercer at Red Bank in October 1777. With his father's assistance and influence, Stuart was able to purchase his commission as Lieutenant-Colonel of the 26th Foot on 26 October 1777. In mid-November, Major-General James Robertson wrote to Lord Bute, informing him that, agreeable to Bute's request, his son had been promoted. "He gets reputation faster than rank; he seeks this by courting danger where it may be found, and he goes out of his way to find it," Robertson noted. "I presume to give you this trouble

8 Don N. Hagist, *General Orders, Rhode Island, December 1776-January 1778* (Bowie, Md: Heritage Books, 2001), p. 36. Professor Gregory Urwin email to Hattendorf, 1 April 2024, including information from Stephen Gilbert.

9 Professor Gregory Urwin email to Hattendorf, 1 April 2024, including information from Stephen Gilbert on Redcoat Images, No. 1,245 (Revisited and Addendum).

10 Stuart-Wortley. *A Prime Minister and His Son*, pp. 116–117, and Anonymous. *New Records of the American Revolution*, p. 47: Stuart to Lord Bute, 27 October 1777.

that by your authority he may be forbidden, but on proper occasions, exposing a life which, if preserved, may be useful to his King and country, and do honour to his family.[11] With the British Army's success and the Delaware River opened to British shipping, Stuart returned to New York in November. In mid-December 1777, Stuart was granted permission to return home on leave. He embarked for England in H.M. frigate *Brilliant*, along with Lord Cornwallis and several other British officers.

Stuart's period of leave allowed him to strike back at Howe by providing the British ministry with a report criticizing the commander-in-chief's conduct of the war in North America. He attended the House of Commons, but there is no record of his making any public speeches in the House. Stuart's letters, however, indicate that he was angered by the Opposition's speeches on the war in America. On 20 April 1778, he married Anne Louisa (1757–1841), the second daughter of Lord Vere Bertie, of Branston, Lincolnshire, who had served as M.P. for Boston in 1741–1754. Louisa's paternal grandfather was Robert, 1st Duke of Ancaster. In mid-May 1778, Stuart embarked in the transport ship *Lyoness* to return to his post with Clinton in New York.[12]

On Stuart's return to America in the early summer of 1778, he found that General Sir William Howe had resigned and gone home, leaving Sir Henry Clinton as commander-in-chief, North America. At this point, Clinton appears to have made Stuart a close confidant. On 7 July, Stuart sailed from New York and arrived at Newport, Rhode Island, the following day, where he found General Pigot's army in good spirits. His diary and letters, printed in this volume, record his activities and observations about military operations in Rhode Island until he returned to New York on 8 September 1778, following the Battle of Rhode Island. In November 1778 Stuart, in poor health, went home on

---

11 Stuart-Wortley. *A Prime Minister and His Son*, p. 117, and Anonymous. *New Records of the American Revolution*, p. 47: Robertson to Lord Bute, 13 November 1777.

12 Stuart-Wortley. *A Prime Minister and His Son*, pp. 122–127.

leave entrusted with carrying letters from Clinton to the ministry.[13] He arrived in London on 28 November.

He returned to North America in mid-July 1779 to lead the Cameronians until November, when the regiment was ordered home. On his return home, Stuart's conduct in actively criticizing Howe and the ministry did nothing to help his military career. He was given no further commands during the war, although made a brevet colonel on 20 November 1782. On 29 July 29, 1784, he gave up his commission in the 26th Foot and exchanged it to become lieutenant-colonel of the 101st Foot stationed in the East Indies. He never joined them but went onto the half-pay after the regiment disbanded on 24 December 1785.[14]

Stuart traveled in Europe on half-pay, envying the attractions of a diplomatic post, but his lack of patience, fiery temper, and egotistical pride made him unlikely to succeed. He continued to represent Bossiney in the House of Commons until 1790, when he vacated the position in favor of his younger brother James. In 1790, he went on to serve Ayr Burghs in Scotland, another constituency associated with family interests, until May 1794, when he vacated the seat in favor of the Duke of Argyll's candidate. Two years later in 1796, he became Member for Poole, Dorset, close to Stuart's residence at High Cliff House, which he had inherited on his father's death in 1792. With the outbreak of war with Revolutionary France in 1793, George III agreed to allowing Stuart to return to active service. Promoted to Major-General on 12 October 1793, he was ordered in April 1794 to command the British Army sent to Corsica to expel the French from that Mediterranean island. At about

13 Anonymous. *New Records of the American Revolution*, p, xxiv; Stuart-Wortley, ed. *A Prime Minister and His Son*, p. 128.

14 Professor Gregory Urwin email to Hattendorf, 1 April 2024 on Redcoat Images No. 1,245 (Revisited and Addendum); Dr. John Houlding's Database of Georgian Officers.

the same time, he also went from half-pay to Colonel of the 68th Foot on 24 October 1794, then Colonel of the 26th Foot on 25 March 1795.[15]

Stuart's army effectively cleared the French from most of the island, then laid siege to the northern city of Calvi with the support of the British Mediterranean fleet under Admiral Samuel Lord Hood. The captain in the fleet who worked closely with Stuart at Calvi was Captain Horatio Nelson, commanding HMS *Agamemnon*. Nelson lost his right eye in the action, when a French canon ball throw up some rocks in his face.[16] After quarrels with Admiral Lord Hood and the British viceroy in Corsica, Sir Gilbert Elliott, over the campaign, Stuart resigned his command in February 1795.

Nearly two years later in January 1797, the government ordered Stuart to Portugal to assist in defending against a French invasion. His troops included several regiments of foreign recruits, who were known for their poor discipline and lack of military spirit. Stuart turned them into a very effective fighting force. In June 1798, the ministry called Stuart back to London to confer with Prime Minister William Pitt on the practicality of recapturing for Britain the Spanish island of Minorca in the Mediterranean's Balearic Islands. Although outnumbered with his army of 3,000 men, Stuart took the island in November 1798 without casualties. For this achievement, the king created him Knight of the Bath and

15 Professor Gregory Urwin email to Hattendorf, 1 April 2024 on Redcoat Images No. 1,245 (Revisited and Addendum); Dr. John Houlding Database of Georgian officers.

16 Stuart-Wortley. *A Prime Minister and His Son*, pp. 239–258; For a detailed history of the Corsican campaign, see Desmond Gregory, *The Ungovernable Rock: A History of the Anglo-Corsican Kingdom and its Role in British Mediterranean Strategy during the Revolutionary War (1793–1797)*. (Cranbury, NJ: Associated University Presses, 1985). On Nelson, see Roger Knight, *The Pursuit of Victory: The Life and Achievements of Horatio Nelson*. (London: Penguin/Allen Lane, 2005), pp. 170–173.

the ministry named him governor of Minorca. Stuart proved to be an effective administrator implementing good government.[17]

In mid-February 1799, Nelson, now a rear-admiral and famous as the victor of the Battle of the Nile, wrote to Stuart requesting that he urgently send troops to defend Sicily. Without refereeing the matter to London, he acted immediately and personally led two regiments to Messina and made a quick visit to Malta, where the French still held the city of Valetta. He returned briefly to Minorca, but in early May ill health forced him to return to England. On his return to London, the ministry immediately offered him command of an army in the Mediterranean area to co-operate with the Austrians. However, he soon feel into disagreements with the duke of York, the commander in chief of the British Army, over the strength of this force and over the ministry's pledge eventually to permit a Russian force at Malta. As a result, Stuart resigned from this post in April 1800. As the duke of York lamented to his father, King George III: "I can only regret that Sir Charles Stuart's unfortunate jealousy of temper and impatience of control from any superior authority preclude the possibility of taking advantage of his otherwise excellent talents."[18]

Nevertheless, the Ministry offered Stuart another assignment, this time as commander-in-chief od British forces in Ireland. General Sir Charles Stuart died on 25 March 1801 at Thatched House, Richmond Park, Surrey, before he could accept the appointment. After nearly 24 years of military service, he was buried in the churchyard of the St Peter's Parish Church, Petersham, Surrey. The village is located today in the modern London Borough of Kingston upon Thames. A monument

17 Stuart-Wortley. *A Prime Minister and His Son*, pp. 259–75, 284–305; For a detailed history, see Desmond Gregory, *Minorca, the Illusory Prize: A History of the British Occupations of Minorca between 1708 and 1802* (Cranbury, NJ: Associated University Presses, 1990), pp. 196–201.

18 Quoted in Desmond Gregory. "Stuart, Sir Charles (1753–1801)," *Oxford Dictionary of National Biography*, from A. Aspinwall, ed., *The Later Correspondence of George III* (Cambridge: Cambridge University Press, 1962–1970), vol. 3, p. 337.

to General Stuart is inside the church. Another monument, a wall plaque by the sculptor Jospeh Nollekens, is in St. Andrews Chapel, Westminster Abbey.[19]

Epilogue. General Stuart left the High Cliff House property to his son, Charles (1779–1845), who had been born during his father's service in North America. Educated at Eton and Christ Church, Oxford, Charles became a diplomat at age 22, the year his father died, and later rose to the peerage in 1828 as 1st Baron Stuart de Rothesay. Carrying his father's name as Sir Charles Stuart between 1812 and 1828, he served twice as ambassador to France, ambassador to Spain during the Peninsular War and finally as ambassador to Russia between 1841 and 1844. At home, the younger Charles set out on a personal mission to develop the property that his grandfather, Lord Bute, had purchased. His father had sold off much of the property and left the manor house, but Rothesay reacquired the land and commissioned the architect William Donthorne to build for him, between 1831 and 1836, a picturesque and romantic Gothic Revival home that he called Highcliffe Castle. He incorporated in it much original medieval stonework recovered from a ruined abbey and "Grand'Maison" in Normandy, France. On completion of the house, he gathered a wide collection of art and extensive library, and family papers, including his grandfather's and father's as well as his own papers.

On Rothesay's death, Highcliffe Castle passed to his wife and then to the youngest of their two daughters as the castle became the scene of much aristocratic and even royal social life. Eventually in 1891, on the death of the youngest daughter who had become Lady Waterford, the house passed to a distant cousin, an army officer, Edward Stuart Wortley, who eventually rose to the rank of Major-General. In the same year that he acquired Highcliffe Castle, Stuart Wortley married Violet Guthrie

19 Westminster Abbey. https://www.westminster-abbey.org/abbey-commemorations/commemorations/sir-charles-stuart

(1866–1953). She became a writer, who wrote several books about the family, using manuscript materials at Highcliffe.

Under the name of The Hon. Mrs. E. Stuart Wortley, C.B.E, she wrote *A Prime Minister and His Son: From the Correspondence of the 3rd Earl of Bute and of Lt.-General The Hon. Sir Charles Stuart, K.B.* (London: Dutton, 1925). She likely sponsored, if not compiled, the anonymous transcription of the same letters, published privately in a limited edition of 250 copies under the title *New Records of the American Revolution: The Letters, Manuscripts and Documents sent by Lieut.-General Sir Charles Stuart, to his Father, the Earl of Bute, 1775–79; also Letters of General Howe, General Clinton, and other Officers to Sir Charles Stuart during the Revolution, 1779–81.* (London: L. Kashnor, 1927).

Edward and Violet Stuart Wortley had three children, Nicholas, Louisia, and Elizabeth. The latter, known as "Bettine," first married Captain Alistair E.G. Grant of the Lancers in 1917.[20] In the following year she appeared, using her maiden name, along with other aristocratic women in D. W. Griffith's 1918 patriotic silent film, "The Great Love."[21] Later, in 1928, she married Montagu Henry Edmund Cecil Towneley-Bertie, Lord Norreys. That same year he succeeded to his grandfather's titles and became 8th Earl of Abingdon.

A twentieth-century note written on the 1778 Charles Stuart Diary states that "The diary comes from the countess of Abingdon, formerly Miss Bettine Stuart Wortley who is a direct descendant of Sir Charles Stuart. Until it reached this country [i.e., the United States] the diary has always been at Highcliffe." Highcliffe Castle was first leased in 1916, then sold in 1949. The publication of some of the Stuart family papers from Highcliffe in the 1920s may have coincided with sale

20 *Evening Mail*, Wednesday, 15 August 1917, p. 4.

21 *Kentish Independent*, Friday, 19 December 1919, p. 13.

of some of the material. Other Stuart family papers were sold at auction in 1935.[22]

This note in the diary hints at bridging its provenance from Highcliffe to its current home in Barrington, Rhode Island. While no information currently held with the diary provides its exact date of arrival back across the ocean, it is next found in the collection of Barrington resident Frederick S. Peck. Peck, profiled more fully below, actively collected books, fine art, and manuscripts, both in the United States and abroad (through dealers and perhaps in person) in the 1920s and 1930s. Among other themes, he had a particular interest in Rhode Island history. It seems likely that, alerted to its relevance to Rhode Island history, Peck purchased the Charles Stuart Diary at auction or through a dealer in London when other papers of this Stuart family had been made available for purchase, sometime in the period ca. 1925–1935.

Provenance of the Diary in Barrington, Rhode Island.[23] The 1778 Charles Stuart Diary now forms part of the Peck Collection, a subset of the Nicholas Gizzarelli Senior Collection in the Barrington Town Museum, under the stewardship and care of Barrington Preservation Society. The Museum is located in the lower level of the Peck Center, adjacent to Barrington Town Hall, in Barrington's Civic Center Historic District. How the diary came to Barrington relates to two major

22 Some of the papers of Lord Stuart de Rothesay, offered for sale in 1935, were requisitioned by the Foreign Office and are now at The National Archives [U.K.], whose collection notes describe this provenance (The National Archives, FO 342, Stuart de Rothesay Papers, catalogued online at https://discovery.nationalarchives.gov.uk/details/r/C7656): "The manuscript collections of Lord Stuart de Rothesay [were] removed from Highcliffe Castle and announced for sale at auction in November 1935 on behalf of the Earl of Abingdon. The records that now form this series were claimed as the property of the Crown by the Foreign Office, by which department it was thereupon bound." In 1981–2, a further collection of Rothesay's papers were purchased at auction by the National Library of Scotland.

23 Nathaniel L. Taylor and Maria Bruce contributed this section for the Barrington Preservation Society.

figures in Barrington's historical community: Frederick Stanhope Peck (1868–1947) and Nicholas Gizzarelli, Sr. (1912–2003).

Frederick Stanhope Peck was one of Barrington's leading citizens. His ancestor Joseph Peck came to Hingham, Massachusetts, in 1638, moving to Rehoboth, in Plymouth Colony, by 1645. Two of Joseph's sons were among the first English settlers in Swansea, Barrington's mother town, incorporated in 1667. The land on which the Pecks settled was part of the Sowams Purchase, lands purchased in 1653 from Osamequin (also known as Massasoit), Sachem of the Pokanoket Tribe, leader of what would subsequently be known as the Wampanoag Federation. To this day, some Peck descendants remain in Barrington, constituted as a daughter town of Swansea in 1717.[24]

During the nineteenth century, one branch of the Peck family prospered. Asa Peck (1812–1890) and his son Leander Remington Peck (1843–1909) built a successful textile firm. Together they maintained their ancestral Peck family lands in northern Barrington. Re-imagined as a gentleman's farm, they named the property Osamequin Farm, in recognition of their heritage, extending almost as far back as the Sowams Purchase of 1653.

The Pecks were generous benefactors to the Town of Barrington. Sarah (Gould) Peck, widow of Leander, donated the land and funds to construct and furnish a high school in her husband's memory. Located next to Barrington Town Hall, the Leander R. Peck Memorial School opened in September 1917 and operated until 1979. Extensively renovated and re-purposed in the early 1980s, the building re-opened in 1984 as the Peck Center. Continuing its legacy of community service, the Peck Center now serves as home to the Barrington Public Library

24 Peck's lineage was published in The Order of Founders and Patriots of America, *Register of Lineages. 1896–1993* (Baltimore: Gateway Press, 1994), vol. 2, *nos. 1526–2814*, pp. 1147–1148.

and the Barrington Town Museum, as well as other community-based services and organizations.

Frederick Stanhope Peck. Leander's son, Frederick Stanhope Peck (1868–1947), continued to build and diversify the family businesses. In addition to textile manufacturing, he also served as a director of the Industrial Trust Bank. Frederick continued his family tradition of political involvement with election to the Barrington Town Council in 1909, and as a Representative to the Rhode Island General Assembly from 1911 to 1926. Peck served as Chairman of the General Assembly's Finance Committee from 1915 to 1926, then served as State Commissioner of Finance until 1935. An active member of the Republican Party, Frederick Peck was a Republican Party National Committeeman from 1918 to 1932, serving as National Chairman for several years.

In 1907, Frederick Peck completed the first phase of construction of a new house, built on ancestral Peck lands, called Belton Court. The original house had massive, rusticated granite masonry, elaborate woodwork, a central stair hall, and beaten copper lanterns, reflecting the influence of the Arts and Crafts movement. In the 1920s, Peck expanded Belton Court into a larger compound, grander and more medieval in style. By then he had become a serious collector of rare books and manuscripts, as well as decorative and fine arts. Belton Court now included a library and public rooms on a grand scale, a perfect backdrop to showcase his book, manuscript, and fine art collections.

Frederick S. Peck actively collected manuscripts of artistic and

Frederick Stanhope Peck, undated photograph.
Courtesy of the Barrington Preservation Society.

historical importance, including a collection of presidential autographs. In 1927, he first published *George Washington Letters: from the Collection of Frederick S. Peck, Belton Court, Barrington, R.I.,25* facsimiles of four letters in his collection. Five years later, in 1932, he published a new and expanded edition.[26] Peck also collected manuscripts of importance to Rhode Island history; one manuscript in his collection was published in 1943 as *The Providence Oath of Allegiance and Its Signers, 1651–2,*[27] in the series of publications of the Rhode Island Society of Colonial Wars, while Peck was serving as the Society's Governor in 1943–44. Peck was active in other hereditary societies in Rhode Island, including Sons of the American Revolution, Order of the Founders and Patriots of America, and Society of Mayflower Descendants.

By the 1940s, offshoring in the textile industry began to strain the Peck family fortunes. In 1944, Frederick S. Peck sold some of his library at auction and donated a significant collection of Rhode Island historical manuscripts to the Rhode Island Historical Society.[28] Many other manuscripts including the presidential autograph collection were auctioned off early in 1947, shortly before his death, and the bulk of the art and antiques furnishing Belton Court were sold at auction after his death later

---

25 Washington, George, and Frederick S. (Frederick Stanhope) Peck. *George Washington Letters: From the Collection of Frederick S. Peck, Belton Court, Barrington, R.I.* ([Providence?]: s.n., 1927).

26 Washington, George, and Frederick Stanhope Peck. *Autograph Letters of George Washington: From the Collection of Frederick S. Peck, Belton Court, Barrington, R.I.* ([Providence?]: s.n., 1932).

27 Richard LeBaron Bowen, *The Providence Oath of Allegiance and Its Signers, 1651–2,* Contributed by Frederick Stanhope Peck. Publication number 35. (Providence: Rhode Island Society of Colonial Wars, 1943).

28 Classified at RIHS as the Frederick Peck Collection, MSS 9005.

in 1947.[29] Peck bequeathed Belton Court itself to a Providence hospital, but it was deemed unsuitable for medical purposes. It was soon sold and began its afterlife as a campus of several different colleges.

Four boxes of historical manuscripts collected by Peck were not sold or donated to the Rhode Island Historical Society, but remained in the possession of his only child, Helen Peck (1895–1989). These relate primarily to Barrington's history, and consist of eighteenth and early nineteenth century letters, deeds, writs, tax lists, and town and private fiscal accounts. Since several of Frederick Peck's ancestors were Barrington town officers during this period, some of these manuscripts may have been handed down as family possessions from the time of their creation, while others were presumably acquired by purchase, including the 1778 Charles Stuart Diary, as described above.

While three of the four boxes contain single-leaf documents arranged and annotated chronologically, the fourth box contains miscellaneous larger documents, including multi-page bound manuscripts. The 1778 Charles Stuart Diary is housed in this box. Recognizing the significance of this set of material to the history of Barrington, Helen Peck donated this collection, along with several additional boxes of Peck family papers and twentieth-century memorabilia, to Barrington Historian Nicholas Gizzarelli, Sr.[30]

Nicholas Gizzarelli, Sr. Recognized as one of Barrington's most notable historians, native son and life-long resident, Nicholas Gizzarelli, Sr., spent decades amassing one of Barrington's most broad-based

---

29 See for example, Peck, Frederick S. (Frederick Stanhope). *Rare American Historical Autographs and a Few Very Rare Books ... the Collection of Frederick S. Peck ...* (Philadelphia: N.p., 1947); Peck, Frederick S. (Frederick Stanhope). *Rare First Editions of English and American Authors, Autographs, Letters and Manuscripts ... Selections from the Library of the Late Frederick S. Peck ... given to Brown University to Be Sold for the Benefit of the Library ... Public Auction Sale, April 21 ... April 22.* (New York: Parke-Bernet Galleries, Inc., 1947).

30 Where it now forms the Peck Collection, a subset of the Nicholas Gizzarelli Sr. Collection.

and significant historical collections. In 1982 and 1983, just prior to the opening of the newly renovated and re-purposed Peck Center, Mr. Gizzarelli donated the bulk of his collection to the Town Museum, providing the foundation for the Museum's new permanent home in the building's lower level. The Nicholas Gizzarelli Sr. Collection, including the Peck Collection,[31] is the largest single subset of the Town Museum's collections. Even today, as efforts to classify the Nicholas Gizzarelli Sr. Collection progress, and its scope and importance emerge, both scholars and students alike continue to benefit from this collection.

Working almost entirely alone, Mr. Gizzarelli researched, collected, compiled, and composed tens of thousands of pages of documents, as well as an assortment of photographs and memorabilia. Ranging from government records to personal diaries and covering almost every imaginable subject pertaining to Barrington's past, the bulk of the collection is comprised of paper documents. For more than a half century, Nick's prolific and impassioned researching and collecting meshed well with his affable ability to cultivate trusted relationships, allowing him to become a sought-out and well-respected conduit for his collecting efforts.

Nicholas Gizzarelli, Sr., 1981
Courtesy of the Barrington Preservation Society.

During the early 1980's, Mr. Gizzarelli published and distributed fourteen volumes of Barrington history. From works detailing Barrington's brick industry, and its Italian American community, with its

31 The Peck Collection at Barrington Preservation Society Museum is classed as Accession no. 2020.17; the Nicholas Gizzarelli Sr. Collection is still in the process of formal description and classification.

origins rooted in the Italian Diaspora, Nick's research extended seamlessly back in Barrington's history, from genealogical compilations of colonial families, to annotated transcriptions of Barrington's earliest vital records, to the Civil War letters of Sergeant Edward A. Goodwin.

In 1990, Mr. Gizzarelli established and endowed the Nicholas Gizzarelli, Sr., Scholarship Fund at Barrington High School and donated educational materials and historical artifacts to the Barrington School Department. In October 1990, and again in March 1993, Nicholas Gizzarelli, Sr., was recognized as Town Historian by the Barrington Town Council. In a more recent acknowledgement of his contributions, Nick was elected as one of the inaugural members of Barrington's Heritage Hall of Fame in 2020. As a testament to his keen insights, focus, and dedication, Nicholas Gizzarelli, Sr.'s legacy continues to be visible through the countless events, exhibits and publications that have, and will use his materials, as the basis of their research pursuits. It is through the efforts and dedication of these two men, Frederick Stanhope Peck and Nicholas Gizzarelli, Sr., that the Sir Charles Stuart *Diary of the Battle of Rhode Island* is now offered for publication.

Select Bibliography for Further Reading

Anonymous. *New Records of the American Revolution: The Letters, Manuscripts and Documents sent by Lieut.-General Sir Charles Stuart, to his Father, the Earl of Bute, 1775–79; also Letters of General Howe, General Clinton, and other Officers to Sir Charles Stuart during the Revolution, 1779–81* (London: Privately Printed by L. Kashnor, 1927).

Conley, Patrick T., *The Battle of Rhode Island August 29, 1778: A Victory for the Patriots* (Providence: Rhode Island Publication Society, 2005).

Gregory, Desmond. "Stuart, Sir Charles (1753–1801)," *Oxford Dictionary of National Biography*. https://doi.org/10.1093/ref:odnb/26697

Gregory, Desmond. *Minorca, the Illusory Prize: A History of the British Occupations of Minorca between 1708 and 1802* (Cranbury, NJ: Associated University Presses, 1990).

Gregory, Desmond. *The Ungovernable Rock: A History of the Anglo-Corsican Kingdom and its Role in British Mediterranean Strategy during the Revolutionary War (1793–1797)* (Cranbury, NJ: Associated University Presses, 1985).

Hagist, Don N. *General Orders, Rhode Island, December 1776-January 1778* (Bowie, Md: Heritage Books, 2001).

Hattendorf, John B., ed. *Mary Gould Almy's Journal During the Siege at Newport, Rhode Island, 29 July to 24 August 1778* (Pennsauken Township, NJ: Bookbaby for the Rhode Island Society Sons of the Revolution, 2018).

Hattendorf, John B., ed. *The Battle of Rhode Island in 1778: The Official View as Reported in The London Gazette. A Facsimile with an Introduction and an Annotated Transcription* (Middletown, RI: The Stone Tower Press for the Rhode Island Sons of the Revolution, 2021).

Mackesy, Piers. *The War for America, 1775–1783* (London: Longmans Green, & Co. 1964).

Mackenzie, Frederick. *Diary of Frederick Mackenzie, Giving a Daily Narrative of His Military Service as an Officer of the Regiment of Royal Welch Fusiliers during the Years 1775–1781 in Massachusetts, Rhode Island, and New York*. Two volumes (Cambridge: Harvard University Press, 1930).

McBurney, Christian M. *The Rhode Island Campaign: The First French and American Operation in the Revolutionary War* (Yardley, PA: Westholme Publishing, 2011).

Schroder, Walter K. *The Hessian Occupation of Newport and Rhode Island, 1776–1779* (NP: Heritage Books, 2005).

Sedgwick, Romney ed. *Letters from George III to Lord Bute, 1756–1766* (London: Macmillan and Co., Ltd, 1939).

Stuart-Wortley, The Hon. Mrs. E. *A Prime Minister and His Son: From the Correspondence of the 3rd Earl of Bute and of Lt.-General The Hon. Sir Charles Stuart, K.B.* (London: Dutton, 1925).

Willcox, William B., ed. *The American Rebellion: Sir Henry Clinton's Narrative of his Campaigns, 1775–1782, with an Appendix of Documents* (New Haven: Yale University Press, 1954).

Editorial Principles

The editor has attempted to preserve the original capitalization, spelling and punctuation of the original handwritten text, but has italicized all ship's names for clarity and uniformity. Footnotes have been used at the bottom of the transcribed text to clarify and explain references in the text that might not be readily understood by the reader. In addition, the editor has provided brief biographical sketches of individuals mentioned in the text.

PART ONE

The Diary

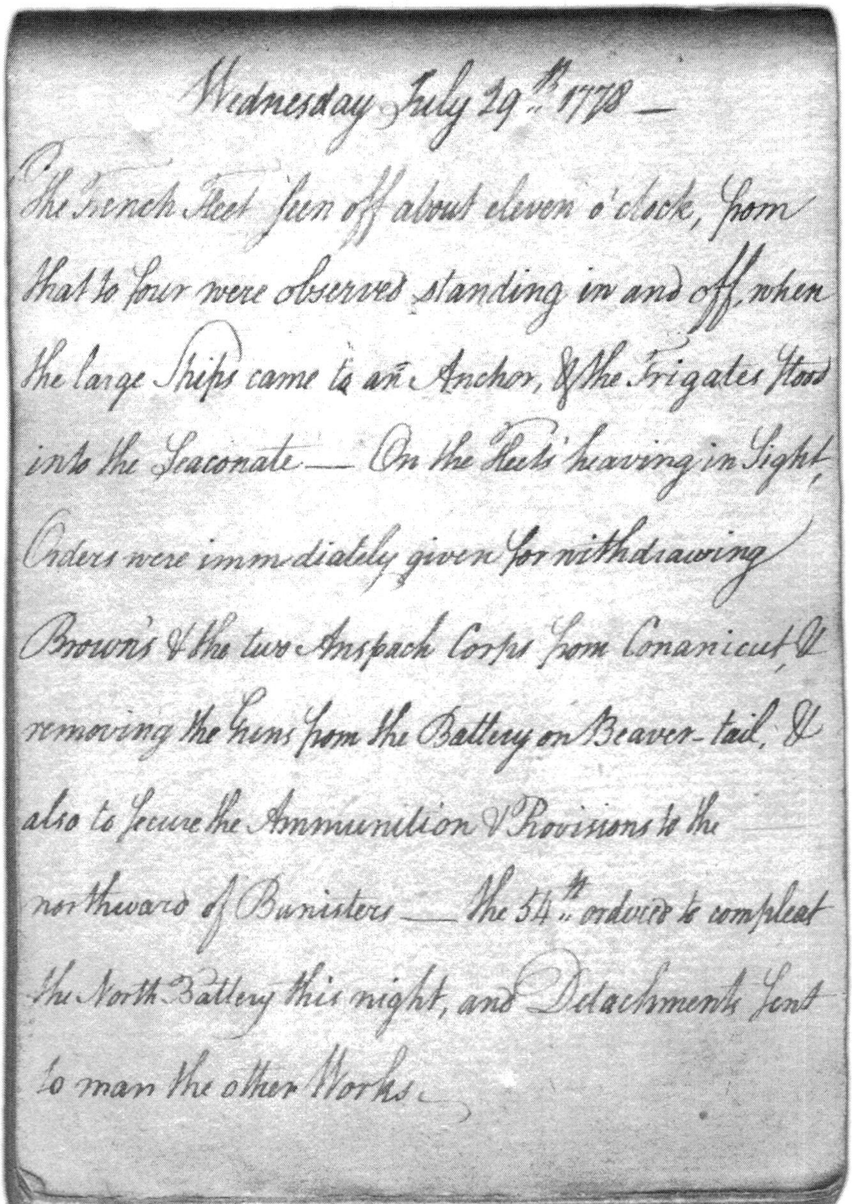

Wednesday July 29th 1778 —

The French Fleet seen off about eleven o'clock, from that to four were observed standing in and off, when the large Ships came to an Anchor, & the Frigates stood into the Seaconate — On the Fleets heaving in Sight, Orders were immediately given for withdrawing Brown's & the two Anspach Corps from Conanicut, & removing the Guns from the Battery on Beaver-tail, & also to secure the Ammunition & Provisions to the northward of Banisters — the 54th ordered to compleat the North Battery this night, and Detachments sent to man the other Works —

Wednesday, July 29th 1778.

The French Fleet[1] seen off about eleven o'clock, from that to four were observed standing in and off, when the large Ships came to an Anchor, & the Frigates Stood into the Seaconate — On the Fleet's heaving in Sight, Orders were immediately given for withdrawing Brown's[2] & the two Anspach Corps from Conanicut,[3] & removing the Guns from the Battery on Beaver-tail, & also to secure the Ammunition & Provisions to the northward of Banisters—The 54th ordered[4] to compleat the North Battery[5] this night, and Detachments Sent to man the other Works.

1 The 17-ship French fleet, commanded by the Comte d'Estaing and consisted of two 80-guns ships of the line, six 74s, three 64s, one 50, and five 26-gun frigates.

2 Captain David Brown's Provincial Militia. Brown was a merchant from Boston who commanded a corps of loyal militia in 1778. Their service seems to have been entirely on Conanicut Island and, after their withdrawal, in defending Newport.

3 The troops of Margrave Friedrich Carl Alexander of Anspach-Bayreuth were the Anspach regiment (commanded by Colonel Fredrich August Valentine Voit von Salzburg) and the Bayreuth Regiment (commanded by Colonel Johann Heinrich Wilhelm von Seybothen).

4 The 54th Regiment of Foot, commanded by Lieutenant-Colonel Andrew Bruce.

5 Now Battery Park on Washington Street, Newport.

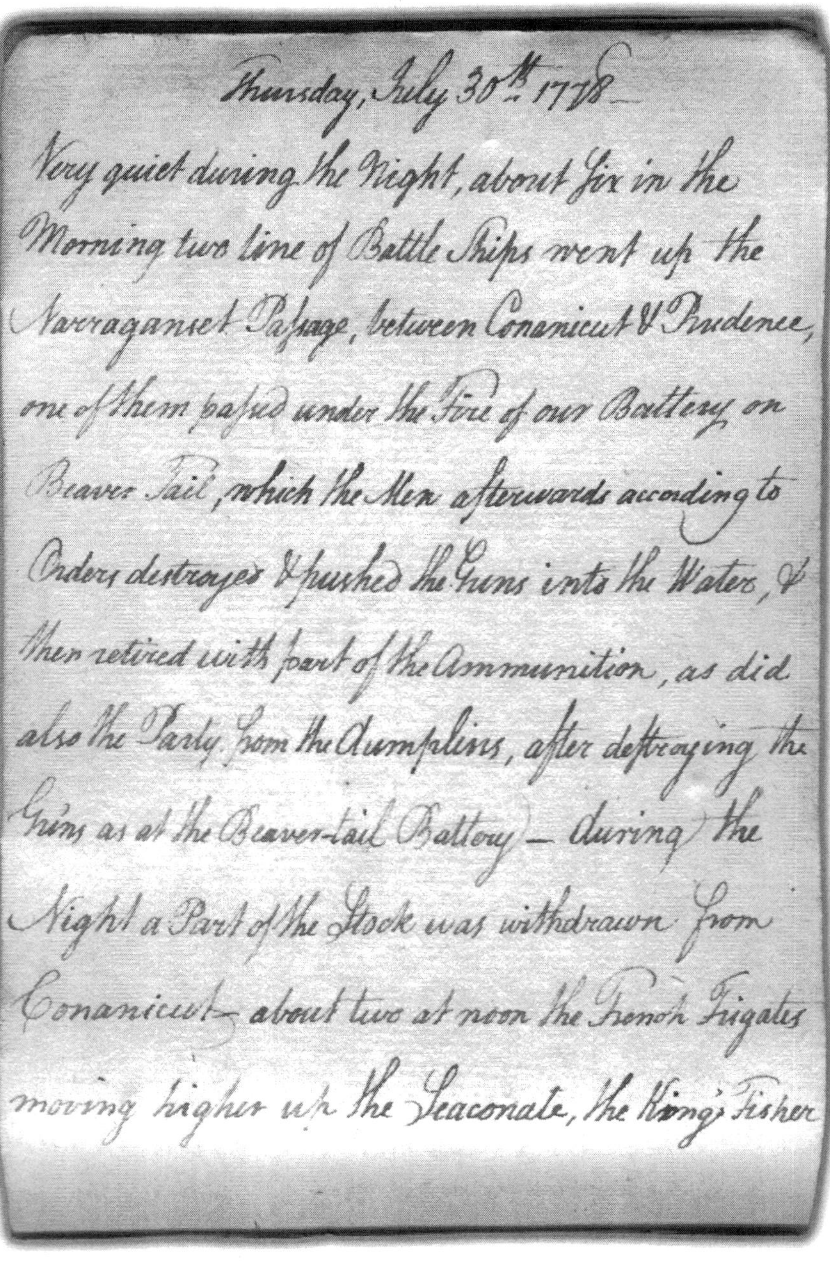

Thursday, July 30th 1778

Very quiet during the Night, about six in the Morning two Line of Battle Ships went up the Narraganset Passage, between Conanicut & Prudence, one of them passed under the Fire of our Battery on Beaver Tail, which the Men afterwards according to Orders destroyed & pushed the Guns into the Water, & then retired with part of the Ammunition, as did also the Party from the Dumplins, after destroying the Guns as at the Beaver-tail Battery — during the Night a Part of the Stock was withdrawn from Conanicut — about two at noon the French Frigates moving higher up the Seaconate, the Kings Fisher

Thursday, July 30th 1778.

Very quiet during the Night, about Six in the Morning two line of Battle Ships went up the Narraganset Passage, between Conanicut & Prudence, one of them passed under the Fire of our Battery on Beaver Tail, which the men afterwards according to Orders destroyed & pushed the Guns into the Water, & then retired with part of the Ammunition, as did also the Party from the Dumplins, after destroying the Guns as at the Beaver-tail Battery — during the Night a Part of the Stock was withdrawn from Conanicut — about two at noon the French Frigates moving higher up the Seaconate, the King's Fisher

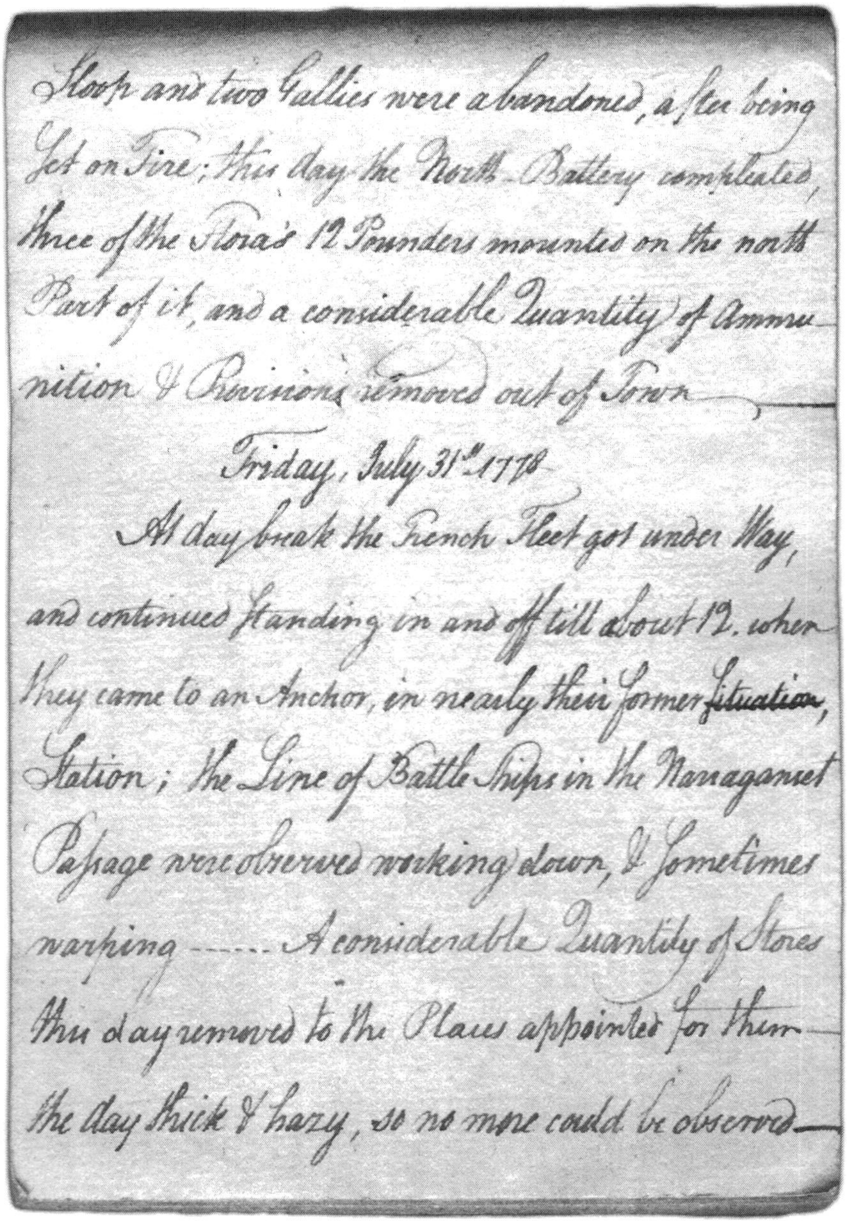

Sloop and two Gallies were abandoned, after being
Set on Fire; this day the North Battery compleated,
three of the Flora's 12 Pounders mounted on the north
Part of it, and a considerable Quantity of Ammu-
nition & Provisions removed out of Town

Friday, July 31st 1778

At day break the French Fleet got under Way,
and continued Standing in and off till about 12, when
they came to an Anchor, in nearly their former ~~Situation~~,
Station; the Line of Battle Ships in the Narraganset
Passage were observed working down, & sometimes
warping ----- A considerable Quantity of Stores
this day removed to the Places appointed for them
the day thick & hazy, so no more could be observed

Sloop[6] and two Gallies[7] were abandoned, after being Set on Fire; this day the North Battery compleated, three of the Flora's[8] 12 Pounders mounted on the north Part of it, and a considerable Quantity of Ammunition & Provisions removed out of town ——

Friday, July 31st 1778.

At daybreak the French Fleet got under Way, and continued Standing in and off till about 12 when they came to an Anchor, in nearly their former ~~situation~~, Station; the Line of Battle Ships in the Narraganset Passage were observed working down, & Sometimes warping ----- A considerable Quantity of Stores this day removed to the Places appointed for them — the day thick & hazy, so no more could be observed. ——

6 The 14-gun, Swan Class ship sloop *Kingfisher* had been built at Chatham dockyard in 1769–70 to the 1766 design of John William. She had a gundeck length of 96 feet, 8 ½ inches; keel length 78 feet 10 ½ inches; breadth 26 feet 10 inches; 302 8/94 tons BM. She was deliberately set on fire, broke from her anchorage, and eventually blew up off High Hill Point, Tiverton, Rhode Island, to avoid capture on 30 July 1778. Her remains have not yet been located.

7 The galleys, *Alarm* and *Spitfire* were sunk on 30 July 1778. *Alarm*, commanded by Lieutenant Philp Auvergne, was set afire and exploded south of McCorrie Point in the Sakonnet River, while *Spitfire*, commanded by Lieutenant James Saumarez, was set on fire and blew up, probably off High Hill Point, Tiverton. Their remains have not been located.

8 The 36-gun frigate *Flora* was commanded by Captain John Brisbane, the senior British naval officer at Newport. Built for the French Navy as the *La Blonde* class frigate *La Vestale* at Le Havre in 1755–56, HMS *Unicorn* took her as a prize on 8 January 1761 off Pointe de Penmarc'h, Brittany. Refitted for service in the Royal Navy at Portsmouth in 1761, she served under the British flag until she was scuttled in Narragansett Bay on 5 August 1788. Salved by the Americans, she was refitted and sold to France in July 1784 as *La Flore Américaine*. She was retaken by the British in 1798, but not brought into the Navy again. She had a gundeck length of 131 foot 7 inches; keel length of 110 feet, 1 ½ inches; breadth 24 feet 6 inches, draft, 10 feet 9 inches; 698 67/98 tons BM. She had a crew of 220 and carried 26 12-pounder guns on the upper deck, 4 six-pounders on the quarterdeck and 2 6- pounders on the fo'c'sle.

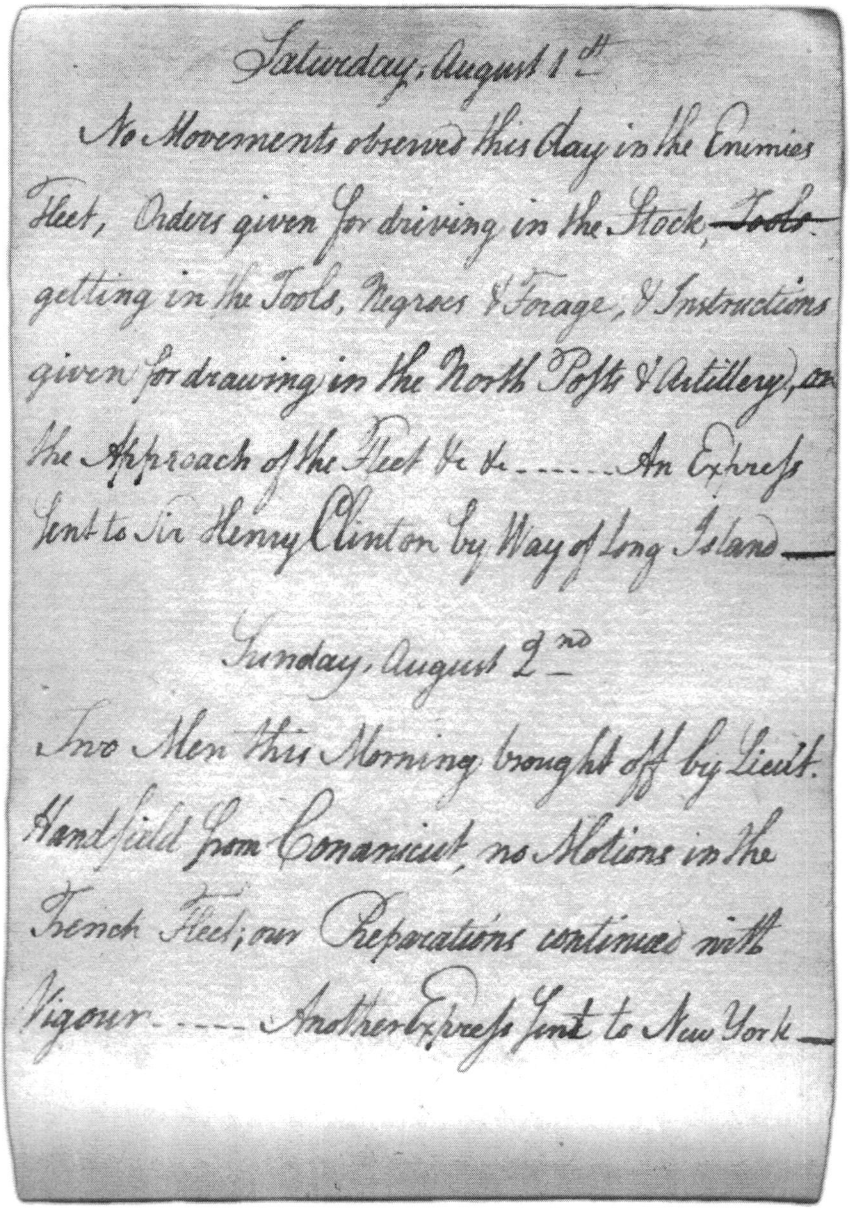

Saturday, August 1st

No Movements observed this day in the Enemies Fleet, Orders given for driving in the Stock, ~~Tools~~ getting in the Tools, Negroes & Forage, & Instructions given for drawing in the North Posts & Artillery), on the Approach of the Fleet &c &c ...... An Express Sent to Sir Henry Clinton by Way of Long Island —

Sunday, August 2nd

Two Men this Morning brought off by Lieut. Handfield from Conanicut, no Motions in the French Fleet; our Preparations continued with Vigour ---- Another Express Sent to New York —

Saturday, August 1st

No Movements observed this day in the Enemies Fleet, Orders given for driving in the Stock, ~~Tools~~ getting in the Tools, Negroes & Forage, & Instructions given for drawing In the North Posts & Artillery, on the Approach of the Fleet &c &c ----- An Express Sent to Sir Henry Clinton by Way of Long Island —

Sunday, August 2nd

Two Men this Morning brought off by Lieut. Handfield[9] from Conanicut, no Motions in the French Fleet; our Preparations continued with Vigour ----- Another Express Sent to New York —

9 Charles Handfield was born in Nova Scotia in 1752. He was one of six sons of John Handfield (born in England in 1693), who was with the 40th Foot in the Canadian Maritimes from 1720 until 1760, when he left the army as the 40th Foot's Lieutenant-Colonel; all six of his sons became officers in the regular army. Charles was commissioned an Ensign in the 22nd Foot in England, 1 October 1769, probably by purchase. He was promoted to Lieutenant in the 22nd Foot in Scotland on 30 September 1772, by purchase, and served as Adjutant, 22 January 1776 to 4 December 1779. He was promoted to Captain-Lieutenant in the 22nd Foot, 15 October 1780 in a free death-vacancy. He became Captain of own Company in the 22nd Foot, 19 March 1789 in a free death-vacancy. Promoted to Lieutenant-Colonel in the new-raising 89th Foot, 3 December 93, he became brevet Colonel 26 January 1797. He resigned from the 89th when he was made Commissary General of Stores on the Dublin staff, 1 March 1799 (to ?). He died as brevet Colonel on 11 April 1831, at Lucan.

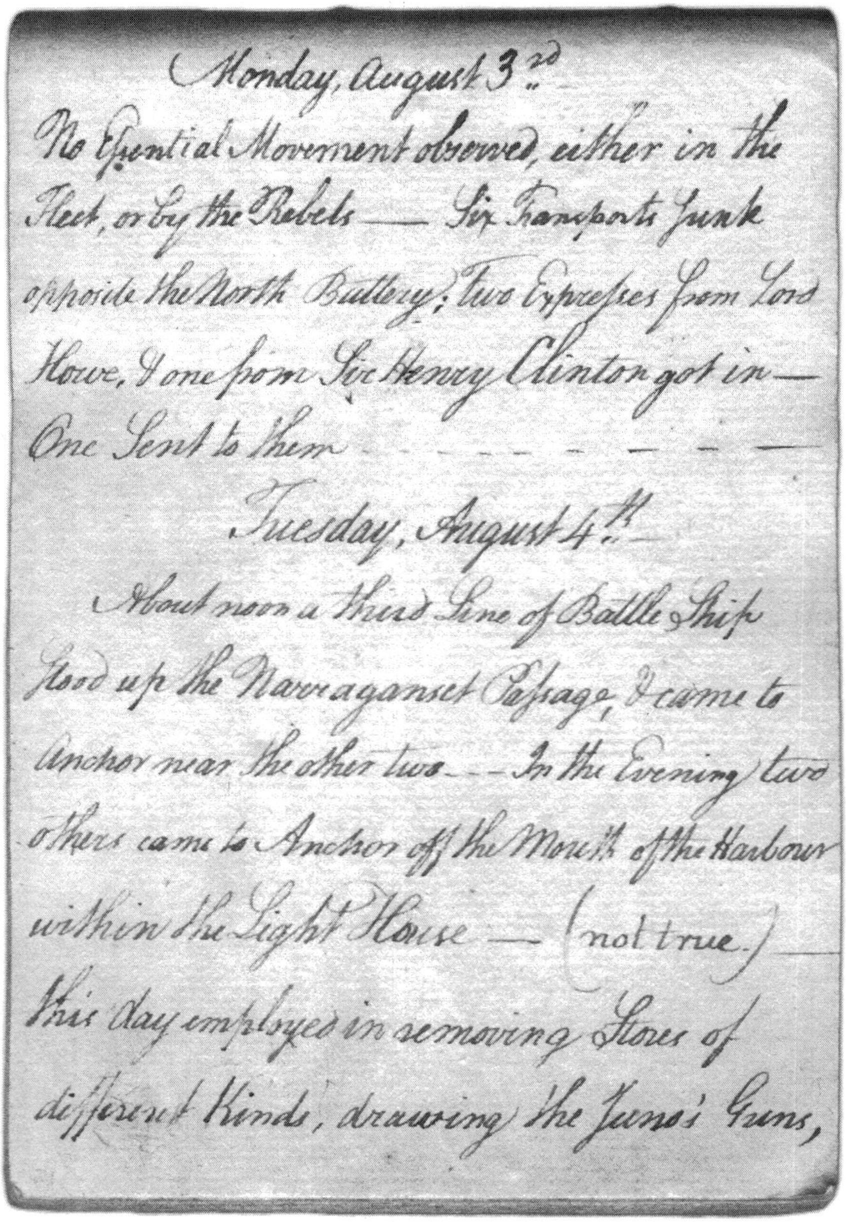

Monday, August 3rd.

No Essential Movement observed, either in the Fleet, or by the Rebels — Six Transports sunk opposite the North Battery; two Expresses from Lord Howe, & one from Sir Henry Clinton got in — One Sent to them - - - - - - - -

Tuesday, August 4th.

About noon a third Line of Battle Ship stood up the Narraganset Passage, & came to Anchor near the other two - - - In the Evening two others came to Anchor off the Mouth of the Harbour within the Light House — (not true.)

This day employed in removing Stores of different Kinds, drawing the Juno's Guns,

Monday, August 3rd

No Essential Movement observed, either in the Fleet, or by the Rebels —— Six Transports sunk opposite the North Battery; Two Expresses from Lord Howe,[10] & one from Sir Henry Clinton got in — One Sent to them ---------------

Tuesday, August 4th

About noon a third Line of Battle Ship Stood up the Narraganset Passage, & came to Anchor near the other two ----- In the Evening two others came to Anchor off the Mouth of the Harbour within the Light House — (not true) — this day employed in removing Stores of different Kinds, drawing the Juno's Guns,[11]

10 Vice-Admiral Lord Viscount Howe

11 On 5 August 1778, the 32-gun Richmond Class frigate *Juno*, commanded by Captain Hugh Dalrymple, was purposely sunk in Coddington Cove. Before being sunk, some of her guns were brought ashore.

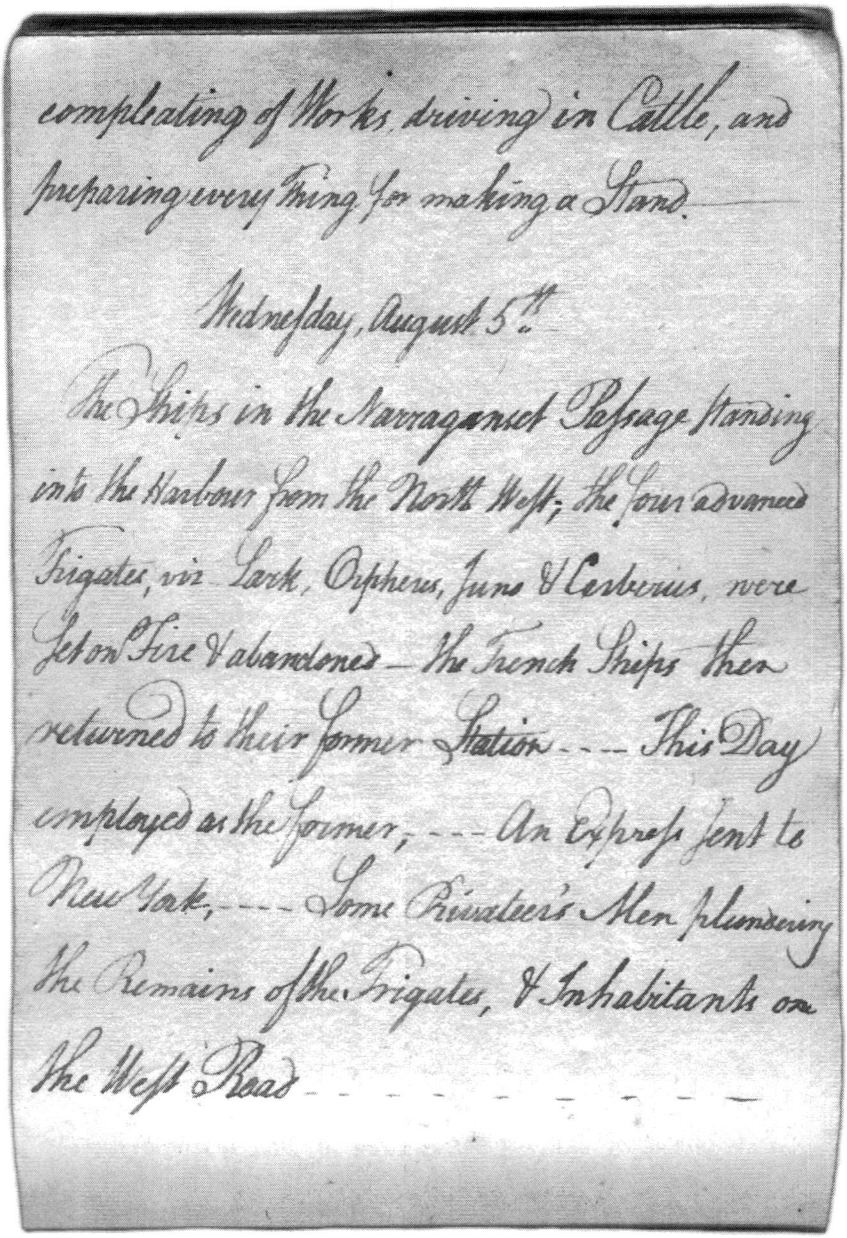

compleating of Works, driving in Cattle, and preparing every Thing for making a Stand.——

Wednesday, August 5th.——

The Ships in the Narraganset Passage standing into the Harbour from the North West; the four advanced Frigates, viz Lark, Orpheus, Juno & Cerberus, were Set on Fire & abandoned — The French Ships then returned to their former Station --- This Day employed as the former; --- An Express sent to New York, ---- Some Privateer's Men plundering the Remains of the Frigates, & Inhabitants on the West Road - - - - - - - - - -

compleating of Works driving in Cattle, and
preparing every Thing for making a Stand. ——

Wednesday, August 5th

The Ships in the Narraganset Passage standing into the Harbour from the North West; the four advanced Frigates, viz. Lark, Orphens, Juno & Cerberus, were Set on Fire & abandoned[12] – the French Ships then returned to their former Station ----- This Day employed as the former, ----- An Express sent to New York, ----- Some Privateer's Men plundering the Remains of the Frigates, & Inhabitants on the West Road --------------

12 The vessels purposely sunk on 5 August 1778 were (1) the 32-gun Richmond Class frigate *Juno*, commanded by Captain Hugh Dalrymple, sunk in Coddington Cove whose remains have not been found but might possibly be the structure found off the east side of Coddington Point in Newport; (2) the 32-gun, modified Lowestoffe Class frigate *Orpheus*, commanded by Captain Charles Hudson, sunk off Melville; (3), the 28-gun, Coventry Class frigate *Cerberus*, commanded by Captain John Symons, sunk about 400 feet off shore south of Carr Point, Middletown; (4) the 32-gun, modified Richmond Class frigate *Lark*, commanded by Captain Richard Smith, sunk off the south side of Arnold's Point in Portsmouth.

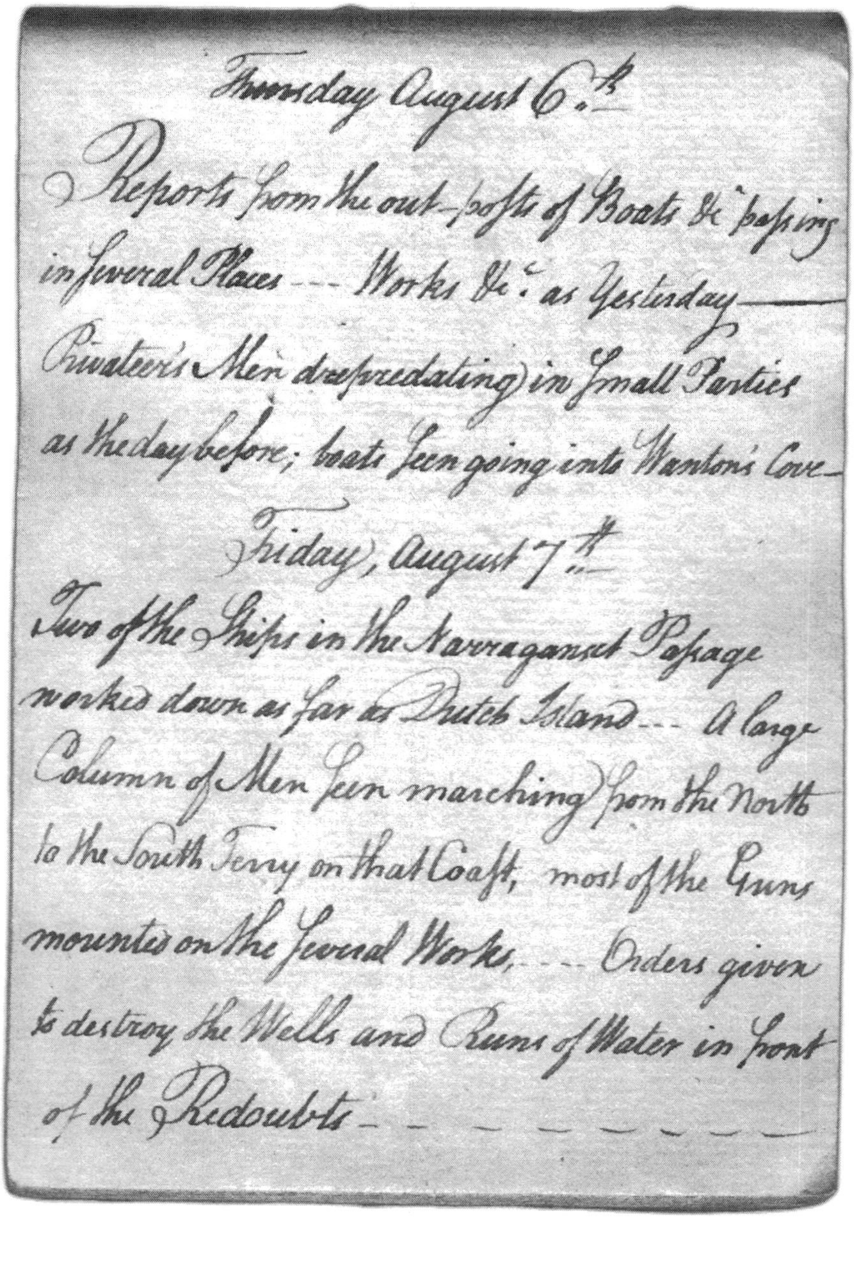

Thursday August 6th.—

Reports from the out-posts of Boats &c passing in several Places --- Works &c. as Yesterday—
Privateer's Men depredating in small Parties as the day before; boats seen going into Wanton's Cove—

Friday, August 7th.—

Two of the Ships in the Narraganset Passage worked down as far as Dutch Island --- A large Column of Men seen marching from the North to the South Ferry on that Coast, most of the Guns mounted on the several Works, --- Orders given to destroy the Wells and Runs of Water in front of the Redoubts— — — — — — — —

Thursday August 6th

Reports from the out-posts of Boats &c passing in Several Places --- Works &c. as Yesterday — Privateer's Men depredating in Small Parties as the day before; boats Seen going into Wanton's Cove.

Friday, August 7th

Two of the Ships in the Narraganset Passage worked down as far as Dutch Island --- A large Column of Men seen marching from the North to the South Ferry on that Coast, most of the Guns mounted on the several Works, --- Orders given to destroy the Wells and Runs of Water in front of the Redoubts ---------------

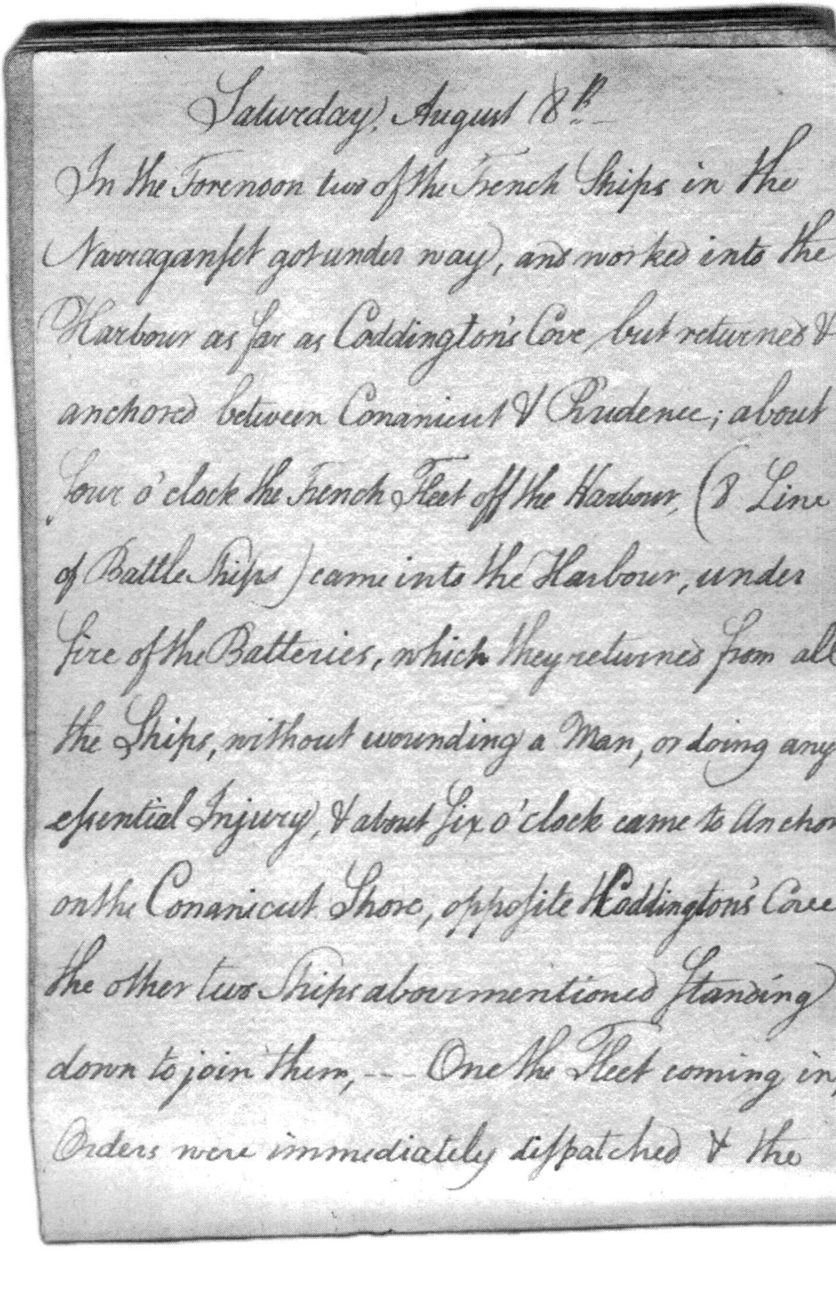

Saturday, August 8th

In the Forenoon two of the French Ships in the Narraganset got under way, and worked into the Harbour as far as Coddington's Cove, but returned & anchored between Conanicut & Prudence; about four o'clock the French Fleet off the Harbour, (8 Line of Battle Ships) came into the Harbour, under fire of the Batteries, which they returned from all the Ships, without wounding a Man, or doing any essential Injury, & about six o'clock came to Anchor on the Conanicut Shore, opposite Coddington's Cove – the other two Ships abovementioned standing down to join them, --- On the Fleet coming in, Orders were immediately dispatched & the

Saturday, August 8th

In the Forenoon two of the French Ships In the Narraganset got under way, and worked into the Harbour as far as Coddington's Cove but returned & anchored between Conanicut & Prudence; about four o'clock the French Fleet off the Harbour, (8 Line of Battle Ships) came into the Harbour, under fire of the Batteries, which they returned from all the Ships, without wounding a Man, or doing any essential Injury, & about Six o'clock came to Anchor on the Conanicut Shore, opposite Coddington's Cove — the other two Ships abovementioned Standing down to join them, --- One the Fleet coming in, Orders were immediately dispatched & the

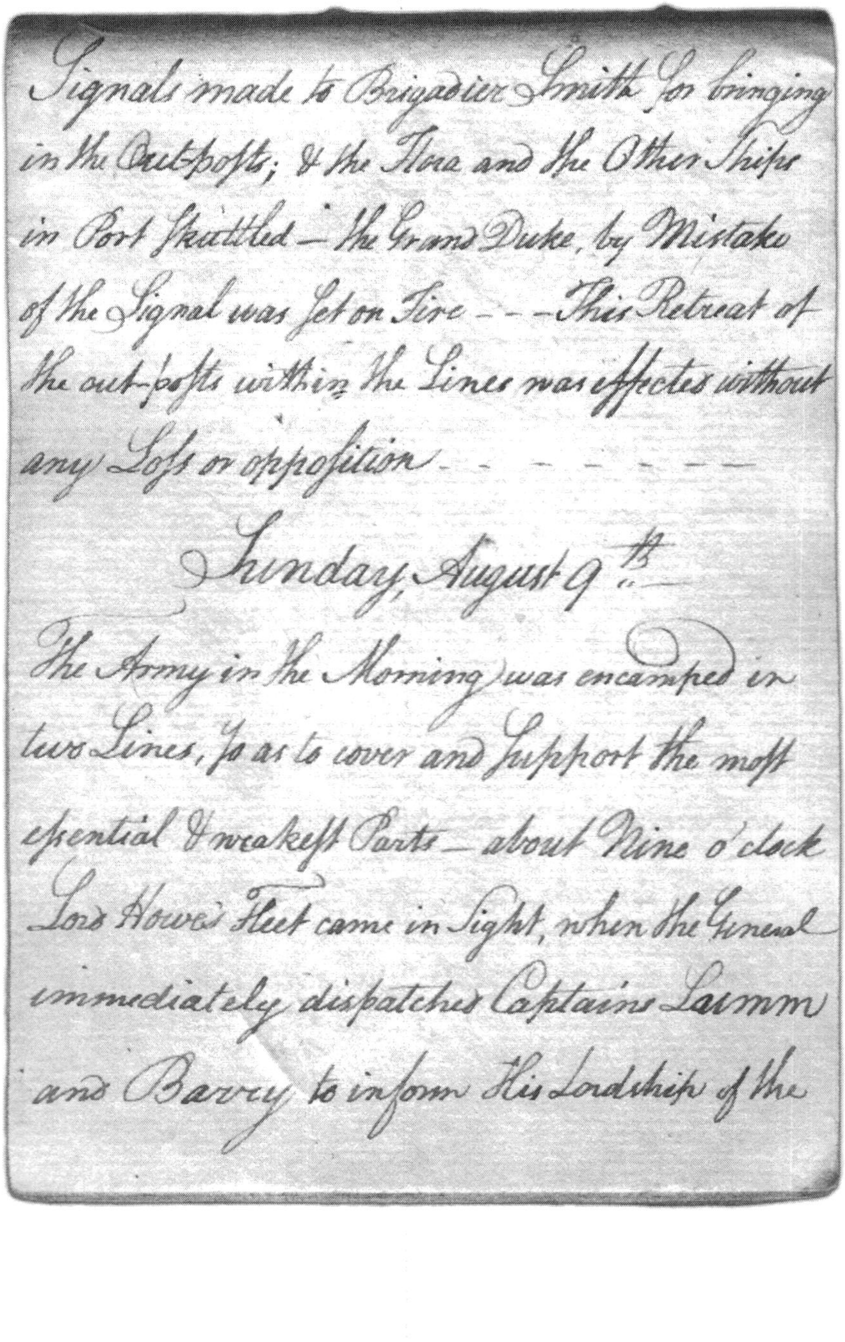

Signals made to Brigadier Smith for bringing in the Out-posts; & the Flora and the Other Ships in Port Scuttled – the Grand Duke, by Mistake of the Signal was set on Fire – – – This Retreat of the out-posts within the Lines was effected without any Loss or opposition – – – – – – –

Sunday, August 9th

The Army in the Morning was encamped in two Lines, so as to cover and support the most essential & weakest Parts – about Nine o'clock Lord Howe's Fleet came in Sight, when the General immediately dispatched Captains Laumm and Barry to inform His Lordship of the

Signals made to Brigadier Smith[13] for bringing in the Out-posts; & the Flora and the Other Ships in Port Skuttled – The Grand Duke,[14] by Mistake of the Signal was set on Fire --- This Retreat of the out-posts within the Lines was effected without any Loss or opposition --------------

Sunday, August 9th

The Army in the Morning was encamped in two Lines, so as to cover and Support the most essential & weakest Parts — About Nine o'clock Lord Howe's Fleet came in Sight, when the General immediately dispatched Captains Laimm[15] and Barry[16] to inform His Lordship of the

13 Francis Smith was born in England in 1723. Smith commanded at Lexington and Concord in April 1775. During the retreat to Boston, he was wounded in the thigh. He was promoted to Brevet Colonel, 8 September 1775 and then obtained a local temporary promotion to Brigadier-General in America, before being promoted to Major-General on 19 February 1779. He became Colonel of the 11th Foot on 10 August 1781. Smith was promoted to Lieutenant-General on 28 September 1787. He died on 7 November 1791 in his home in London, after 50.6 years of service.

14 A storeship.

15 This is likely a reference to Charles Lumm, born in Ireland in 1743. He was promoted to Major in the 44th Foot at New York, 14 April 1779, by purchase. He exchanged to Major in the 79th Foot in Jamaica in a move on paper to facilitate selling out and retired 4 April 1783.

16 See note in Addendum on p. 139.

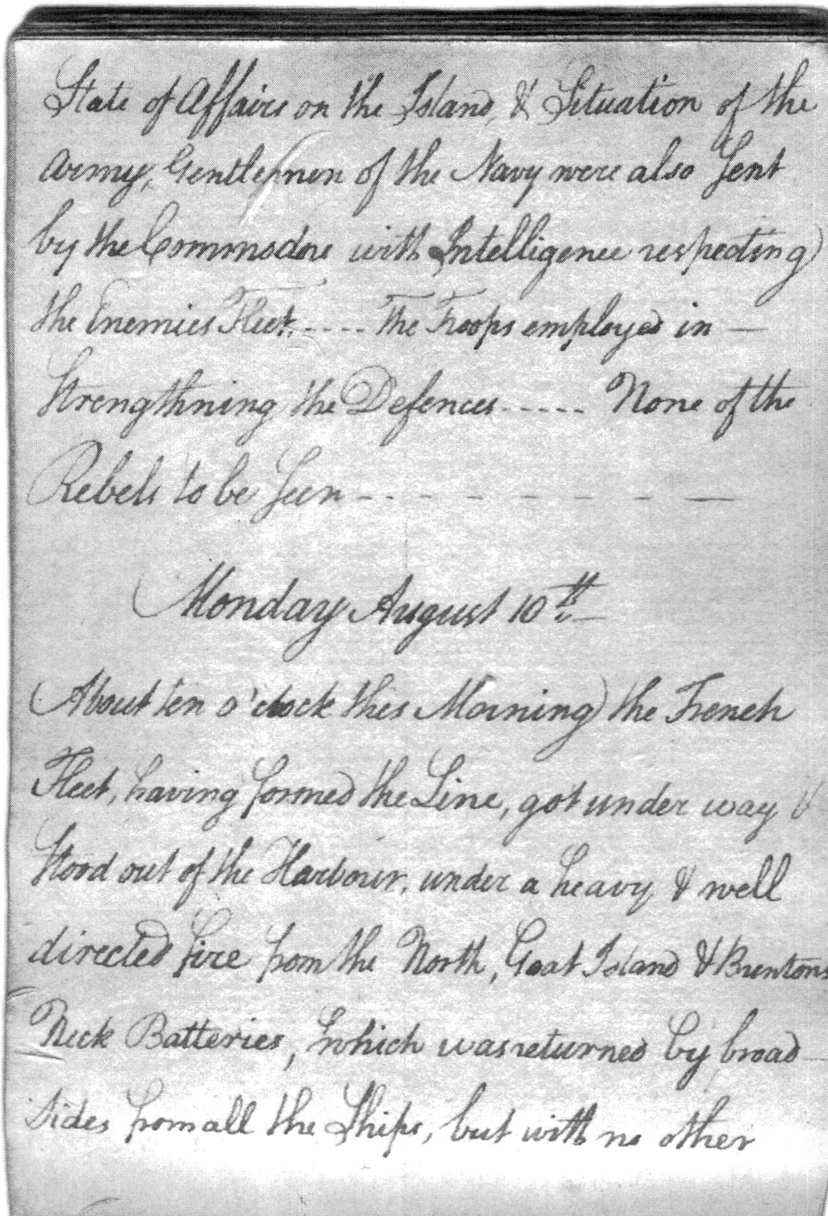

State of Affairs on the Island, & Situation of the Army; Gentlemen of the Navy were also sent by the Commodore with Intelligence respecting the Enemies Fleet---- The Troops employed in Strengthning the Defences---- None of the Rebels to be seen----------

Monday August 10th

About ten o'clock this Morning the French Fleet, having formed the Line, got under way & stood out of the Harbour, under a heavy & well directed fire from the North, Goat Island & Brenton's Neck Batteries, which was returned by broad Sides from all the Ships, but with no other

State of Affairs on the Island, & Situation of the Army, Gentlemen of the Navy were also sent by the Commodore with Intelligence respecting the Enemies Fleet. ----- The Troops employed in — Strengthning the Defences. ----- None of the Rebels to be seen. ---------------

Monday, August 10th

About ten o'clock this Morning the French Fleet, having formed the Line, got under way & Stood out of the Harbour, under a heavy & well directed fire from the North, Goat Island & Brenton's Neck Batteries, which was returned by broad — Sides from all the Ships, but with no other

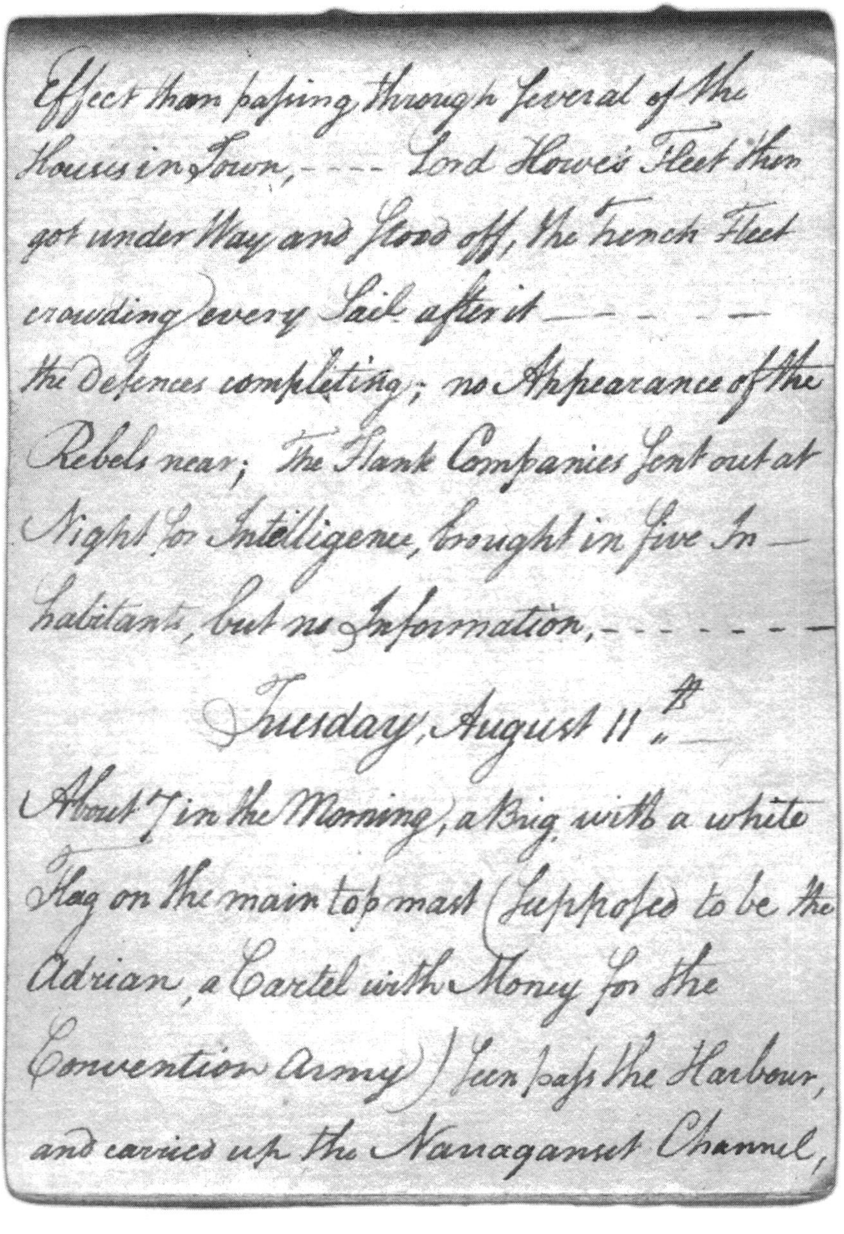

Effect than passing through several of the
Houses in Town, ---- Lord Howe's Fleet then
got under Way and stood off, the French Fleet
crowding every Sail after it ------
the Defences completing; no Appearance of the
Rebels near; the Flank Companies sent out at
Night for Intelligence, brought in five In-
habitants, but no Information, ------

Tuesday, August 11th

About 7 in the Morning, a Brig with a white
Flag on the main top mast (supposed to be the
Adrian, a Cartel with Money for the
Convention Army) seen pass the Harbour,
and carried up the Narraganset Channel,

Effect than passing through several of the Houses in Town, ----- Lord Howe's Fleet then got under Way and Stood off, the French Fleet crowding every Sail after it -----
the defences completing; no Appearance of the Rebels near; the Flank Companies Sent out at Night for Intelligence, brought in five In — habitants, but no Information. ---------------

Tuesday, August 11th

About 7 in the Morning, a Brig, with a white Flag on the main top mast (Supposed to be the Adrian, a Cartel with Money for the Convention Army)[17] seen pass the Harbour, and carried up the Narraganset Channel,

17 The Convention Army was the army of British and allied troops that had surrendered to the Americans and given parole after the battle of Saratoga, 17 October 1777.

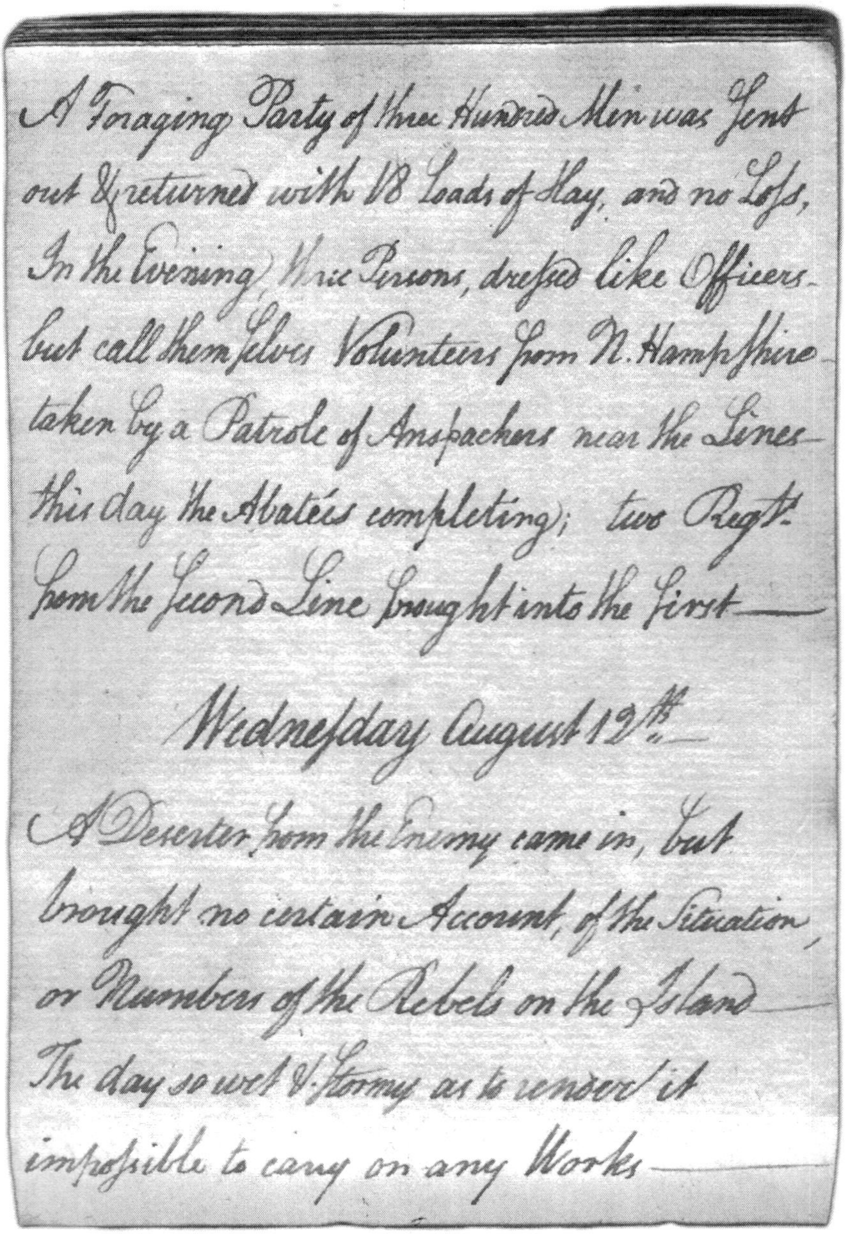

A Foraging Party of three Hundred Men was sent out & returned with 18 Loads of Hay, and no Loss, In the Evening, three Persons, dressed like Officers but call themselves Volunteers from N. Hampshire taken by a Patrole of Anspachers near the Lines this day the Abatées completing; two Regts. from the second Line brought into the First —

Wednesday August 12th. —

A Deserter from the Enemy came in, but brought no certain Account, of the Situation, or Numbers of the Rebels on the Island — The day so wet & stormy as to render it impossible to carry on any Works —

A Foraging Party of three Hundred Men was Sent out & returned with 18 Loads of Hay, and no Loss, In the Evening, three Persons, dressed like Officers, but call themselves Volunteers from N. Hampshire taken by a Patrole of Anspachers near the Lines— this day the Abateis completing; two Regts. from the Second Line brought into the first. —

Wednesday August 12th

A Deserter from the Enemy came in, but brought no certain Account, of the Situation, or Numbers of the Rebels on the Island. — The day so wet & Stormy as to render it impossible to carry on any Works. ———

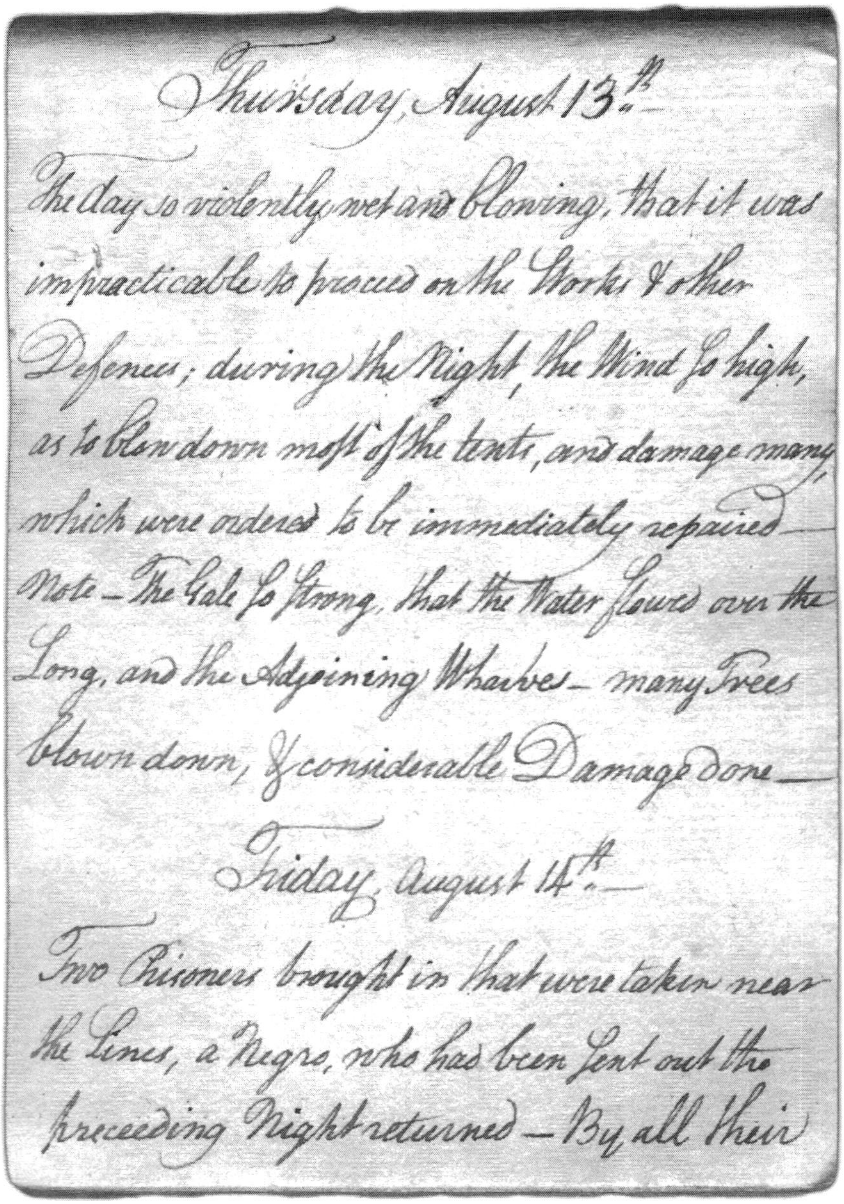
Thursday, August 13th.—
The day so violently wet and blowing, that it was
impracticable to proceed on the Works & other
Defences; during the Night, the Wind so high,
as to blow down most of the tents, and damage many
which were ordered to be immediately repaired—
Note—The Gale so strong, that the Water flowed over the
Long, and the Adjoining Wharves— many Trees
blown down, & considerable Damage done—
Friday, August 14th.—
Two Prisoners brought in that were taken near
the Lines, a Negro, who had been sent out the
preceeding Night returned—By all their

Thursday, August 13th

The day so violently wet and blowing, that it was impracticable to proceed on the Works & other Defenses; during the Night, the Wind so high, as to blow down most of the tents, and damage many, which were ordered to be immediately repaired — Note – The Gale so strong, that the Water flowed over the Long, and the Adjoining Wharves — Many Trees blown down, & considerable Damage done. —

Friday, August 14th

Two Prisoners brought in that were taken near the Lines, a Negro, who had been sent out the preceeding Night returned — By all their

Accounts, the main Body of the Rebels on Quaker
and Windmill Hills, their Numbers uncertain,
--- A body of Rebels observed marching over the
Hills to the Eastward of our Posts, --- the Men em-
ployed in throwing up a Breast work from Green-
end Windmill to Irish's Redoubt, ---- An
Encampment observed near Quaker Hill —

Saturday, August 15th.

The working Parties employed as yesterday —
In the Evening, an Encampment, supposed of
about 3 Regts. seen on the Ground lately
occupied by the 54th. and a Body of Men on
the Heights to the Eastward --- Firing from scattering
Parties

Accounts, the main Body of the Rebels on Quaker and Windmill Hills, their Numbers uncertain, --- A body of Rebels observed marching over the Hills to the Eastward of our Posts --- the Men employed in throwing up a Breast work from Green-end Windmill to Irish's Redoubt, --- An Encampment observed near Quaker Hill. —

Saturday, August 15th

The working Parties employed as yesterday — In the Evening, an Encampment, supposed of about 3 Regts. seen on the Ground lately occupied by the 54th and a Body of Men on the Heights to the Eastward --- Firing from scattering Parties

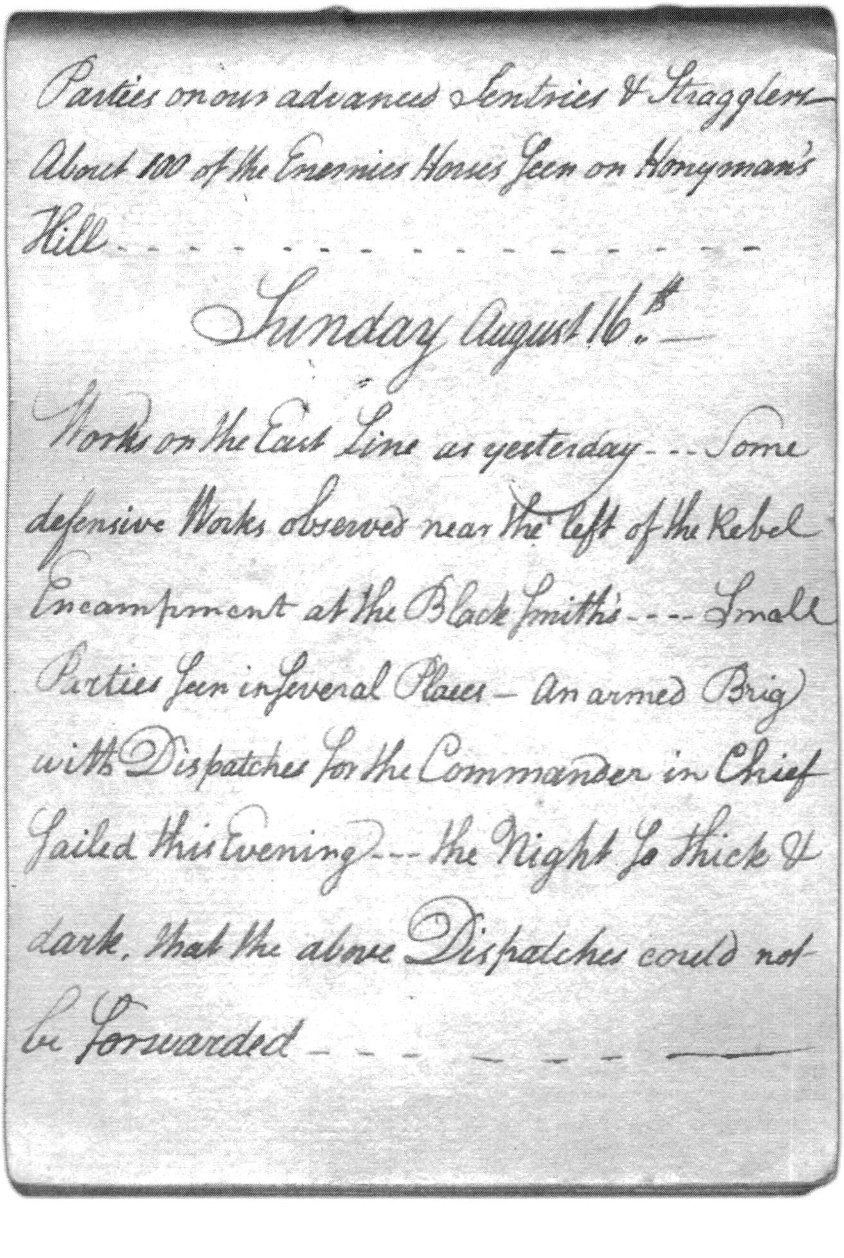

Parties on our advanced Sentries & Stragglers
About 100 of the Enemies Horses seen on Honyman's Hill - - - - - - - - - - - - - -

Sunday August 16th. —

Works on the East Line as yesterday - - - Some defensive Works observed near the left of the Rebel Encampment at the Black Smith's - - - Small Parties seen in several Places — An armed Brig with Dispatches for the Commander in Chief sailed this Evening - - - the Night so thick & dark, that the above Dispatches could not be Forwarded - - - - - - —

Parties on our advanced Sentries & Stragglers —
About 100 of the Enemies Horses seen on Honyman's
Hill --

Sunday, August 16th

Works on the East Line as yesterday —— Some defensive Works observed near the left of the Rebel Encampment at the Black Smith's —— Small Parties seen in several Places — An armed Brig with Dispatches for the Commander in Chief sailed this Evening - - - - - the Night so thick & dark, that the above Dispatches could not be forwarded - - - - - - - - - - -

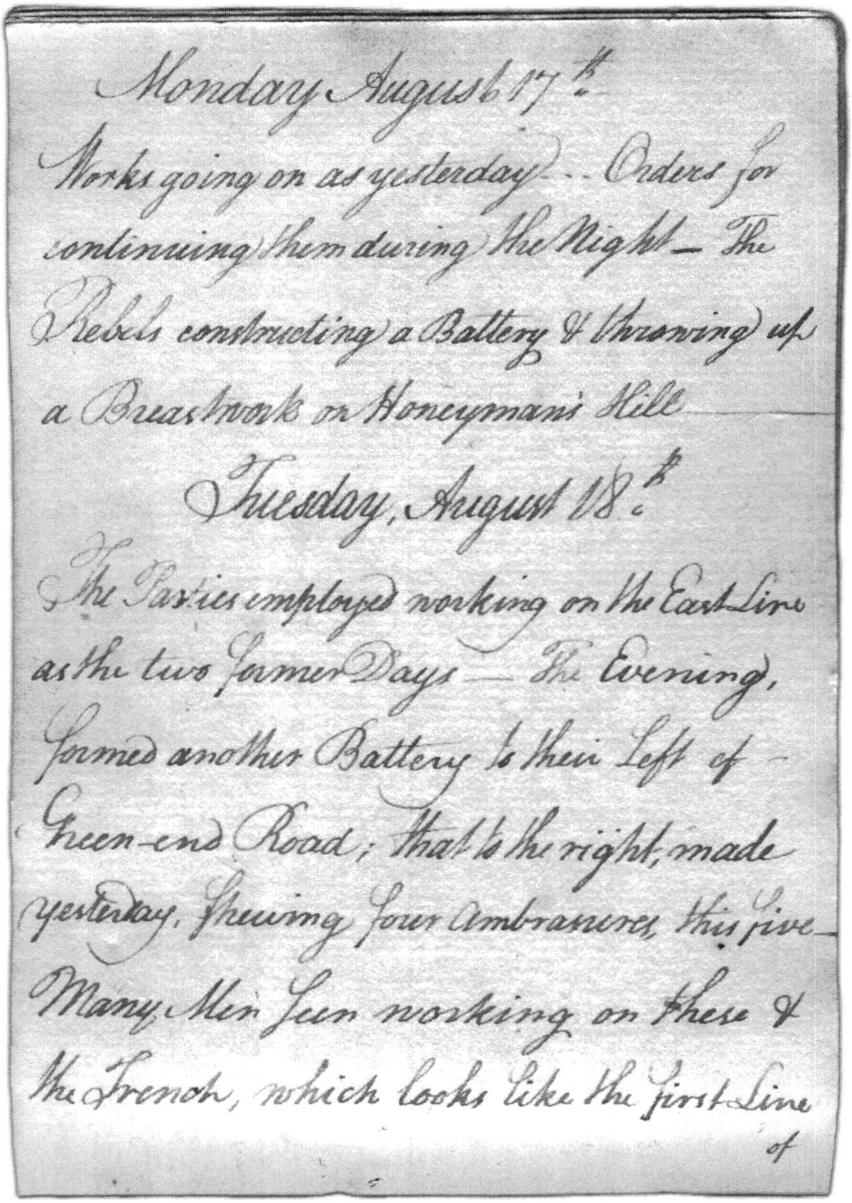

Monday August 17th

Works going on as yesterday — Orders for continuing them during the Night — The Rebels constructing a Battery & throwing up a Breastwork on Honeyman's Hill —

Tuesday, August 18th

The Parties employed working on the East Line as the two former Days — The Evening, formed another Battery to their Left of Green-end Road; that to the right, made yesterday, shewing four Ambrasures, this five — Many Men seen working on these & the French, which looks like the first Line of

Monday, August 17th

Works going on as yesterday — Orders for continuing them during the Night — The Rebels constructing a Battery & throwing up a Breastwork on Honeyman's Hill ————

Tuesday, August 18th

The Parties employed working on the East Line as the two former Day —— The Evening, formed another Battery to their Left of Green-end Road; that to the right, made yesterday, Shewing four Ambrasures, this five —— Many Men seen working on these & the French, which looks like the first Line

of

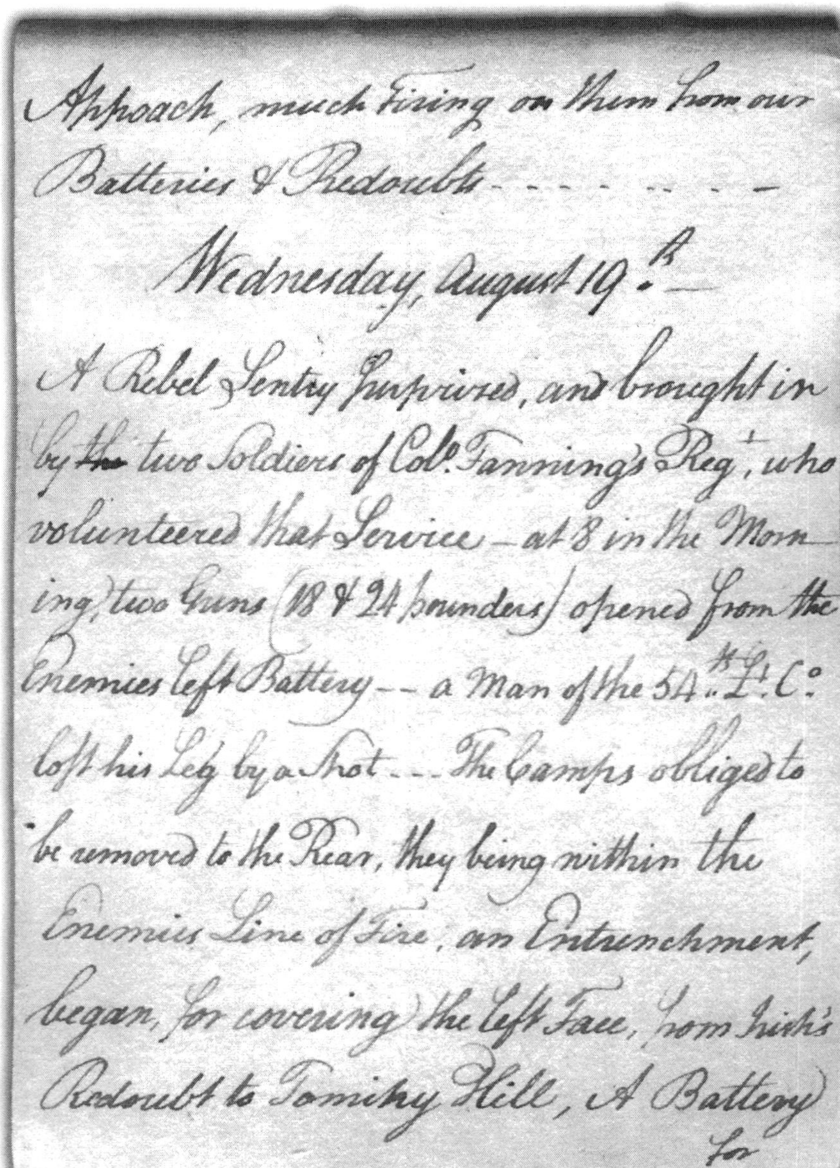
Approach, much Firing on them from our
Batteries & Redoubts - - - - - - - -
Wednesday, August 19th —
A Rebel Sentry Surprised, and brought in
by ~~the~~ two Soldiers of Colo. Fanning's Regt., who
volunteered that Service — at 8 in the Morn-
ing, two Guns (18 & 24 pounders) opened from the
Enemies Left Battery -- a Man of the 54th Lt. Co.
lost his Leg by a Shot ... The Camps obliged to
be removed to the Rear, they being within the
Enemies Line of Fire, an Entrenchment,
began, for covering the Left Face, from Irish's
Redoubt to Tominky Hill, A Battery
for

Approach, much Firing on them from our Batteries & Redoubts ---------------

Wednesday, August 19th

A Rebel Sentry surprised, and brought in by ~~the~~ two Soldiers of Col°. Fanning's Regt.,[18] who volunteered that Service — at 8 in the Morning, two Guns (18 & 24 pounders) opened from the Enemies left Battery --- a Man of the 54th Lt. C°. lost his Leg by a Shot --- The Camps obliged to be removed to the Rear, they being within the Enemies Line of Fire, an Entrenchment, began, for covering the left Face, from Irish's Redoubt to Tominy Hill, A Battery

for

18 This was the loyalist, King's American Regiment. Edmund Fanning was an American from New York, born in 1737. Before the Revolution, he had been Colonel of the Orange County Militia in North Carolina. In December 1776, General Howe appointed him Colonel of the King's American Regiment. He was wounded at Rhode Island in 1778. In 1779, he became Colonel and Commandant of the Rhode Island Loyal Associated Refugees. On Christmas Day 1782, he was reappointed Colonel of the King's American Regiment and went on to Half Pay as a Colonel in 1783. He served as Lieutenant-Governor of Nova Scotia from 24 February 1783 to 20 May 1786, then served as Lieutenant-Governor of St. John's Island (Prince Edward Island) from 4 November 1786 to 10 May 1804 and was Commandant, Island of St John Volunteers of the Royal Canadian Militia from 25 June 1795 to 24 June 1798. He was promoted to Major-General on 12 October 1793, and then to Lieutenant-General on 26 June 1799, and General on 25 April 1808. He died in London on 28 February 1818.

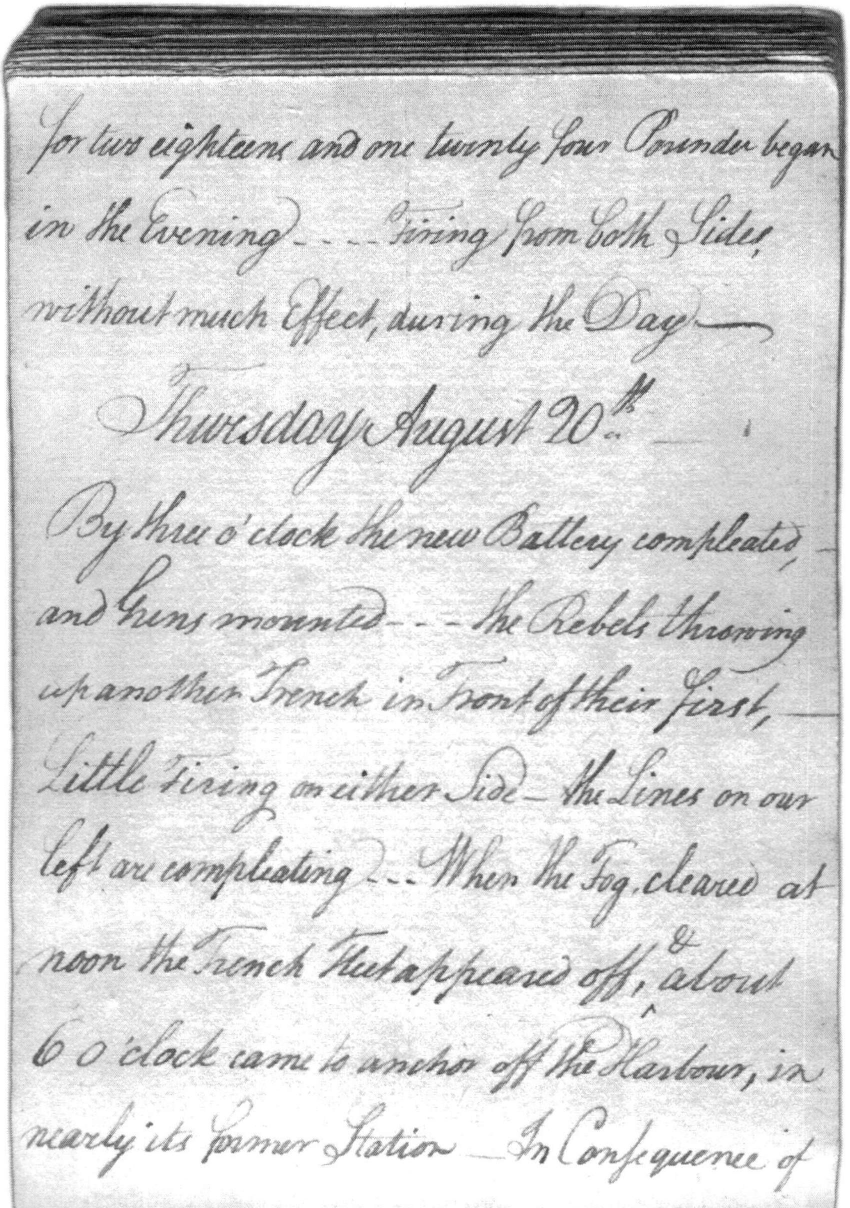

for two eighteens and one twenty four Pounder began in the Evening --- Firing from both Sides, without much Effect, during the Day —

Thursday August 20th.

By three o'clock the new Battery compleated, and Guns mounted --- the Rebels throwing up another Trench in Front of their first, Little Firing on either Side — the Lines on our left are compleating --- When the Fog cleared at noon the French Fleet appeared off, & about 6 o'clock came to anchor off the Harbour, in nearly its former Station — In Consequence of

For two eighteens and one twenty four Pounder began in the Evening —— Firing from both Sides, without much Effect, during the Day ——

Thursday, August 20th

By three o'clock the new Battery compleated, – and Guns mounted — the Rebels throwing up another Trench in Front of their first, — Little Firing on either Side --- the Lines on our left are compleating —— When the Fog cleared at noon the French Fleet appeared off, & about 6 o'clock came to anchor off the Harbour, in nearly its former Station — In Consequence of this Event the Guns remanded from the New Battery to the Works they were brought from, or replaced by others, and the Sea Batteries reinforced by Artillery Men, Sailors &c.ª Some of the French Fleet seen to be under Jury Masts and otherwise disabled — Two more Guns mounted in the Rebels left, and one in their right Battery — — — — — — — — —

Friday, August 21st.

The two Rebel Batteries compleated with Guns, some Firing on both Sides, 4 1/2 inch Shells from the Rebels, supposed to be thrown from 24 Pounders, as most of them fell dead —

this Event the Guns remanded from the New Battery to the Works they were brought from, or replaced by others, and the Sea Batteries reinforced by Artillery Men, Sailors &c. Some of the French Fleet seen to be under Jury Masts and otherwise disabled — Two more Guns mounted in the Rebels left, and one in their right Battery ---------------

Friday, August 21st

The two Rebel Batteries compleated with Guns, some Firing on both Sides, 4 ½ inch Shells from the Rebels, supposed to be thrown from 24 Pounders, as most of them fell dead. —

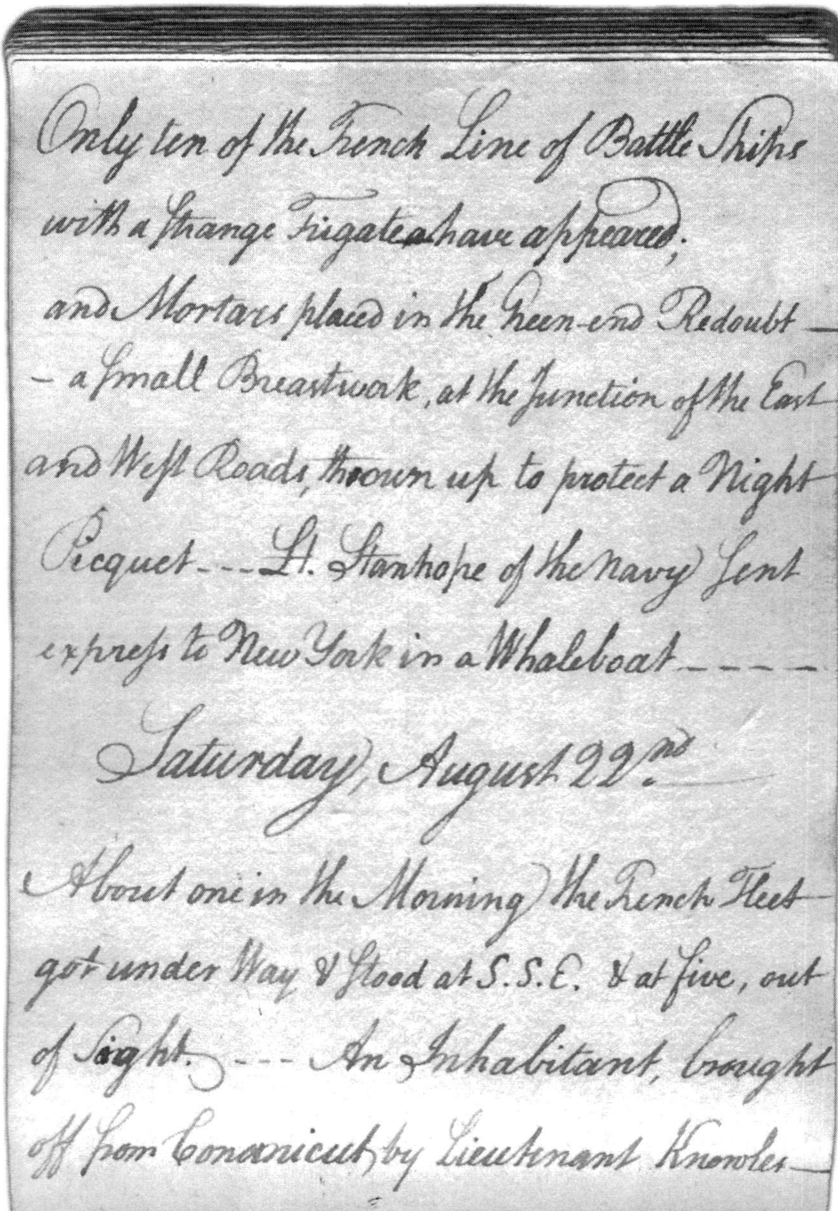

Only ten of the French Line of Battle Ships
with a strange Frigate have appeared;
and Mortars placed in the Green-end Redoubt —
— a small Breastwork, at the Junction of the East
and West Roads, thrown up to protect a Night
Picquet --- Lt. Stanhope of the Navy sent
express to New York in a Whaleboat ----

Saturday, August 22nd.

About one in the Morning the French Fleet
got under Way & stood at S.S.E. & at five, out
of Sight --- An Inhabitant, brought
off from Conanicut by Lieutenant Knowles —

Only ten of the French Line of Battle Ships
with a Strange Frigate have appeared;
and Mortars placed in the Green-end Redoubt —
a small Breastwork, at the Junction of the East
and West Roads, thrown up to protect a Night
Picquet --- Lt. Stanhope[19] of the Navy sent
express to New York in a Whaleboat ----

Saturday, August 22nd

About one in the Morning the French Fleet –
Got under Way & Stood at S.S.E. & at five, out
of Sight. --- An Inhabitant, brought
off from Conanicut by Lieutenant Knowles[20] –

19 Lieutenant The Hon. Henry Edwyn Stanhope, Royal Navy, was an undergraduate at Worcester College, Oxford, when he joined the Navy in May 1768. He served in the 20-gun frigate *Rose* on the North American station. When she was paid off, he returned to Oxford and took his degree. He was commissioned a Lieutenant in 1777 and returned to America. He served in HMS *Nonsuch*, then commanded the armed galley *Pigot* during the raids on Tiverton and Warren, Rhode Island in May 1778. He returned to England in 1780 and was promoted to captain in 1781. In 1783, he carried the definitive peace treaty to New York. On 17 August 1783 in Trinty Church, Newport, he married Margaret, daughter of Francis Malbone. Returning to service he was promoted to Rear Admiral in 1801 and Vice Admiral in 1805. In 1807, he was second in command to Admiral Lord Gambier at Copenhagen, service for which he was created a baronet. He was promoted to admiral in 1812. After retiring, he devoted his time to studying Hebrew and translating the Old Testament. He died in 1814,

20 See note in Addendum on p. 139.

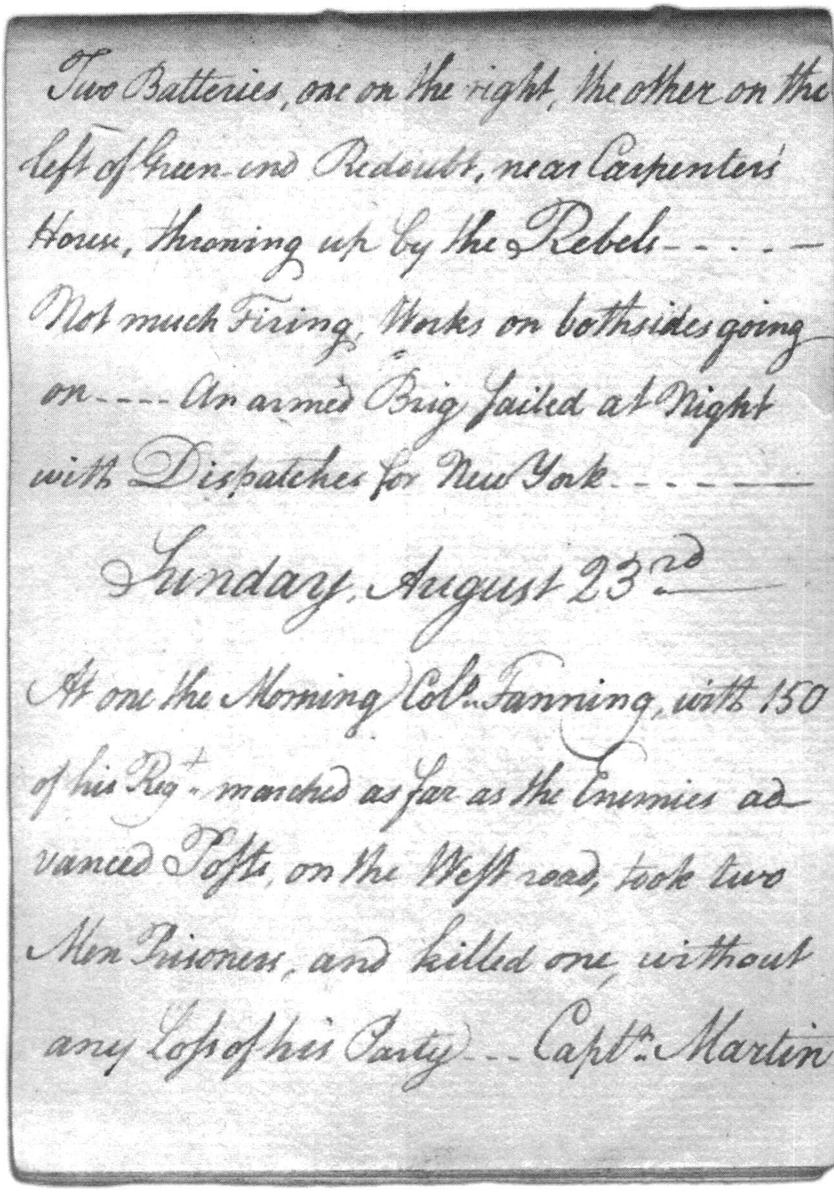

Two Batteries, one on the right, the other on the left of Green-end Redoubt, near Carpenters House, throwing up by the Rebels - - - -

Not much Firing, Works on bothsides going on - - - - An armed Brig sailed at Night with Dispatches for New York - - - - -

Sunday, August 23rd

At one the Morning Col. Fanning, with 150 of his Regt. marched as far as the Enemies advanced Posts, on the West road, took two Men Prisoners, and killed one, without any Loss of his Party - - - Captn. Martin

Two Batteries, one on the right, the other on the left or Green-end Redoubt, near Carpenter's House, throwing up by the Rebels -----
Not much Firing, Works on both sides going on —— An armed Brig sailed at Night with Dispatches for New York -----

Sunday, August 23rd

At one the Morning Col^{o}. Fanning, with 150 of his Reg^{t} marched as far as the Enemies advanced Posts, on the West road, took two Men Prisoners, and killed one, without any Loss of his Party --- Captn. Martin[21]

21 Captain Martin is probably Michael Martyn, born ca. 1767, appointed Deputy Muster Master General of Provincial Forces at Rhode Island, 6 March 1778 to April 1779, and a year later was captain in the Loyal Associated Refugees under George Leonard at Newport. He died on Long Island in 1780.

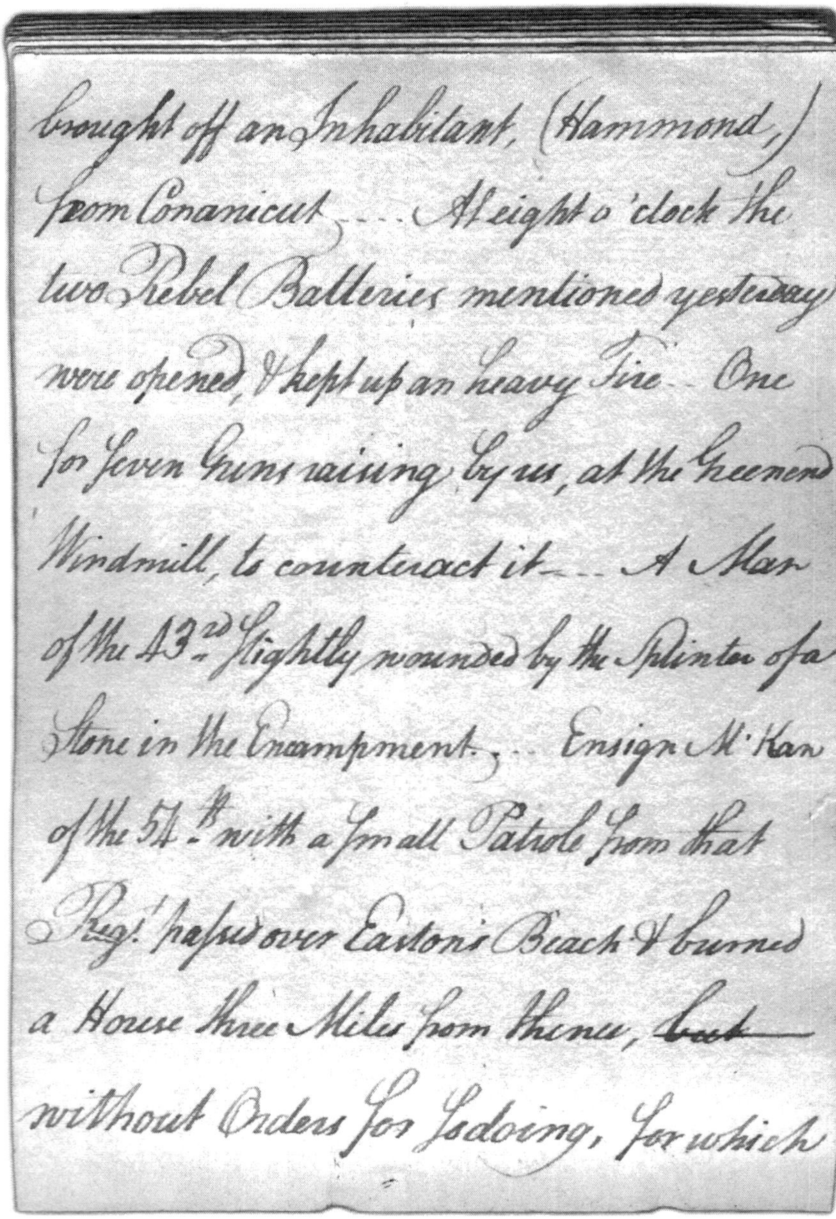

brought off an Inhabitant, (Hammond,)
from Conanicut .... At eight o'clock the
two Rebel Batteries mentioned yesterday
were opened, & kept up an heavy Fire.. One
for seven Guns raising by us, at the Greenends
Windmill, to counteract it.... A Man
of the 43d slightly wounded by the Splinter of a
Stone in the Encampment... Ensign M'Kan
of the 54th with a small Patrole from that
Regt. passed over Easton's Beach & burned
a House three Miles from thence, ~~but~~
without Orders for so doing, for which

brought off an Inhabitant, (Hammond,) from Conanicut ---- At eight o'clock the two Rebel Batteries mentioned yesterday were opened, & kept up an heavy Fire — One for seven Guns raising by us, at the Greenend Windmill, to counteract it ---- A Man of the 43rd slightly wounded by the splinter of a Stone in the Encampment ---- Ensign M'Kan[22] of the 54th with a small Patrole from that Regt. passed over Easton's Beach & burned a House three Miles from thence, ~~but~~ without Orders for so doing, for which

22 William Mecan was born in Ireland, date unknown. He was commissioned an Ensign in the 54th Foot in Ireland, 28 November 1775, by purchase. H was promoted to Lieutenant in the 44th Foot at New York, 26 January 1779 in a free death-vacancy. On 9 September 1779, the 44th Foot, in company with the Hessen-Kassel Regiments of Lossberg and Knyphausen, sailed from New York bound for Quebec. A storm at sea on 13–17 Sept dispersed the fleet with loss, including one of the 44th's transports with three Companies on board. Mecan was one of five (or six) of the 44th's officers lost at sea on 15 September 1779.

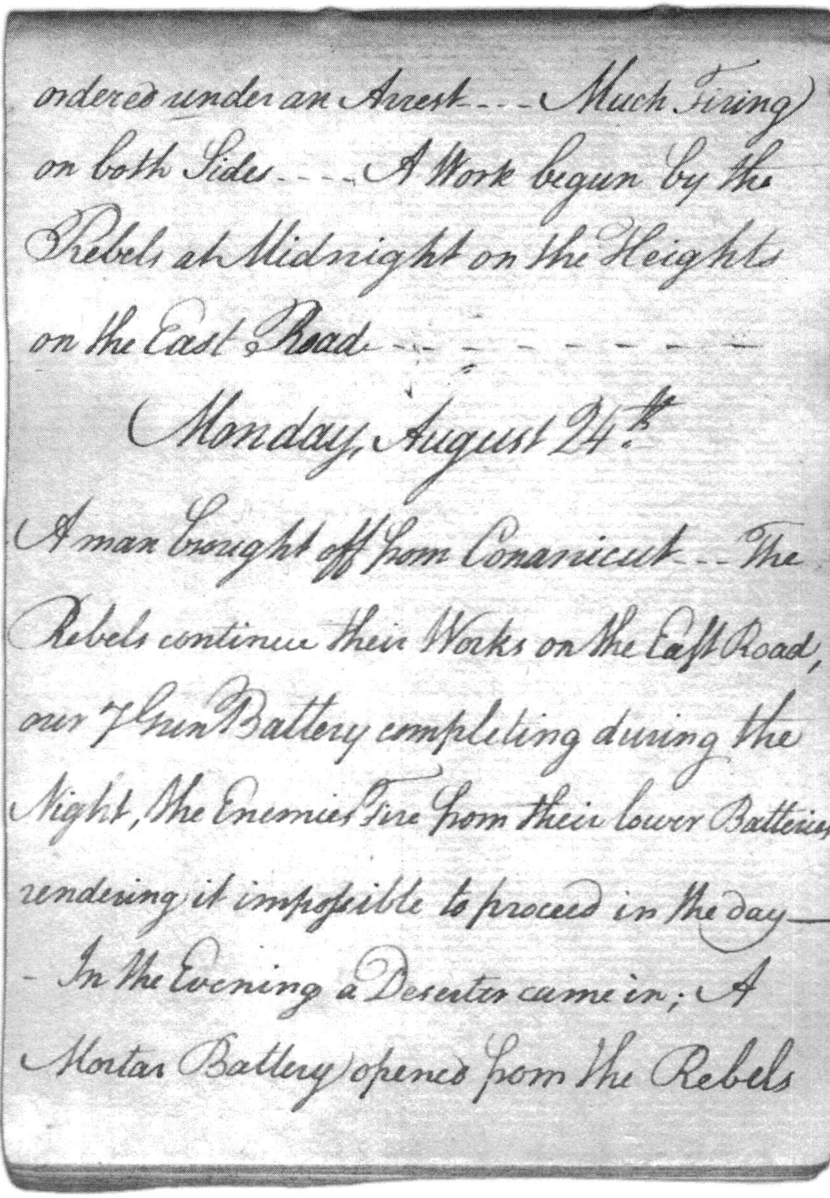
ordered under an Arrest --- Much Firing
on both Sides --- A Work begun by the
Rebels at Midnight on the Heights
on the East Road ----------

Monday, August 24th

A man brought off from Conanicut --- The
Rebels continue their Works on the East Road,
our 7 Gun Battery completing during the
Night, the Enemies Fire from their lower Batteries
rendering it impossible to proceed in the day —
- In the Evening a Deserter came in; A
Mortar Battery opened from the Rebels

ordered under an Arrest ---- Much Firing
on both Sides ---- A Work begun by the
Rebels at Midnight on the Heights
on the East Road ---------------

Monday, August 24th

A man brought off from Conanicut --- The
Rebels continue their Works on the East Road,
our 7 Gun Battery completing during the
Night, the Enemies Fire from their lower Batteries,
rendering it impossible to proceed in the day —
In the Evening a Deserter came in; A
Mortar Battery opened from the Rebels

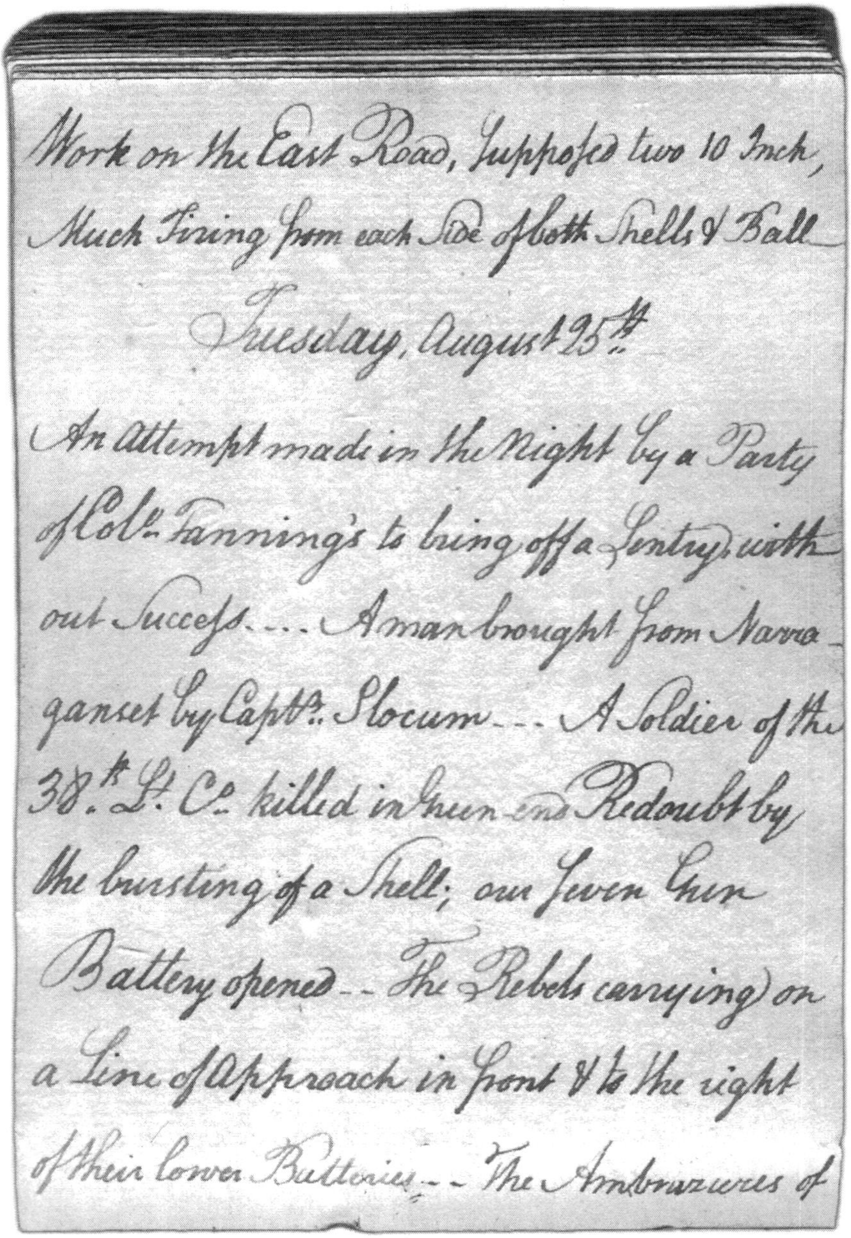
Work on the East Road, Supposed two 10 Inch,
Much Firing from each Side of both Shells & Ball

Tuesday, August 25th.

An Attempt made in the Night by a Party of Colo. Fanning's to bring off a Sentry, without Success.... A man brought from Narragansett by Captn. Slocum... A Soldier of the 38th Lt. Co. killed in Green-end Redoubt by the bursting of a Shell; our Seven Gun Battery opened -- The Rebels carrying on a Line of Approach in front & to the right of their lower Batteries -- The Ambrazures of

Work on the East Road, supposed two 10 Inch,
Much Firing from each Side of both Shells & Ball ——

Tuesday, August 25th

An attempt made in the Night by a Party of Col°. Fanning's to bring off a Sentry with out Success ---- A man brought from Narraganset by Captn. Slocum[23] --- A Soldier of the 38th Lt. C° killed in Green-end Redoubt by the bursting of a Shell; our seven Gun Battery opened -- The Rebels carrying on a Line of Approach in front & to the right of their lower Batteries -- The Ambrazures of

23 Captain Slocum may be Ebenezer Slocum of North Kingston, Rhode Island, born ca. 1755, who in ca. 1777 was given a warrant as a captain to raise a company in the Loyal New Englanders. Warranted as a Lieutenant 30 October 1780 in Hazard's Corps, Volunteer Militia, Suffolk County, New York.

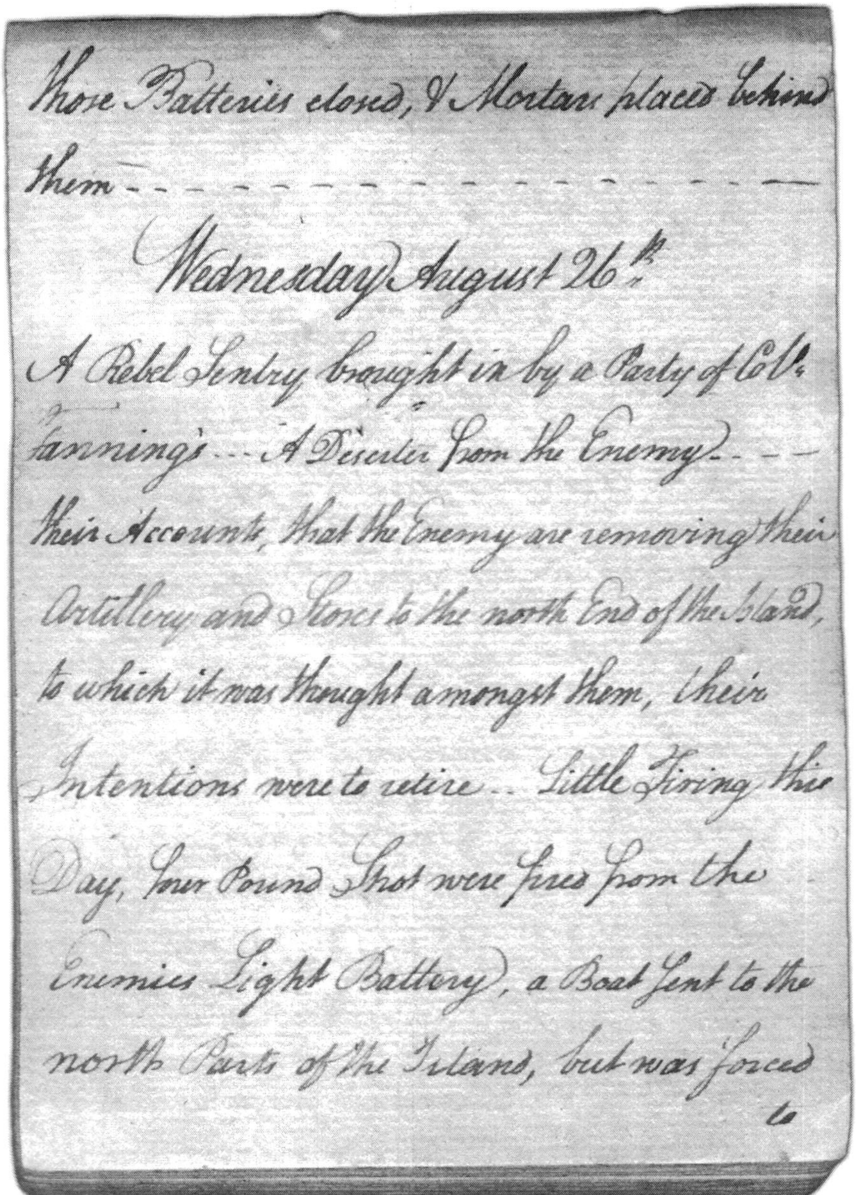
those Batteries closed, & Mortars placed behind
them - - - - - - - - - - - - - - - - - - -

Wednesday August 26th

A Rebel Sentry brought in by a Party of Col.
Fanning's --- A Deserter from the Enemy - - -
their Accounts, that the Enemy are removing their
Artillery and Stores to the north End of the Island,
to which it was thought amongst them, their
Intentions were to retire .. Little Firing this
Day, Four Pound Shot were fired from the
Enemies Light Battery, a Boat sent to the
north Parts of the Island, but was forced
to

those Batteries closed, & Mortars placed behind them --

Wednesday, August 26th

A Rebel Sentry brought in by a Party of Col°. Fanning's --- A Deserter from the Enemy ---- their Accounts, that the Enemy are removing their Artillery and Stores to the north End of the Island, to which it was thought amongst them, their Intentions were to retire -- Little Firing this Day, four Pound Shot were fired from the Enemies Light Battery, a Boat sent to the north Parts of the Island, but was forced

to

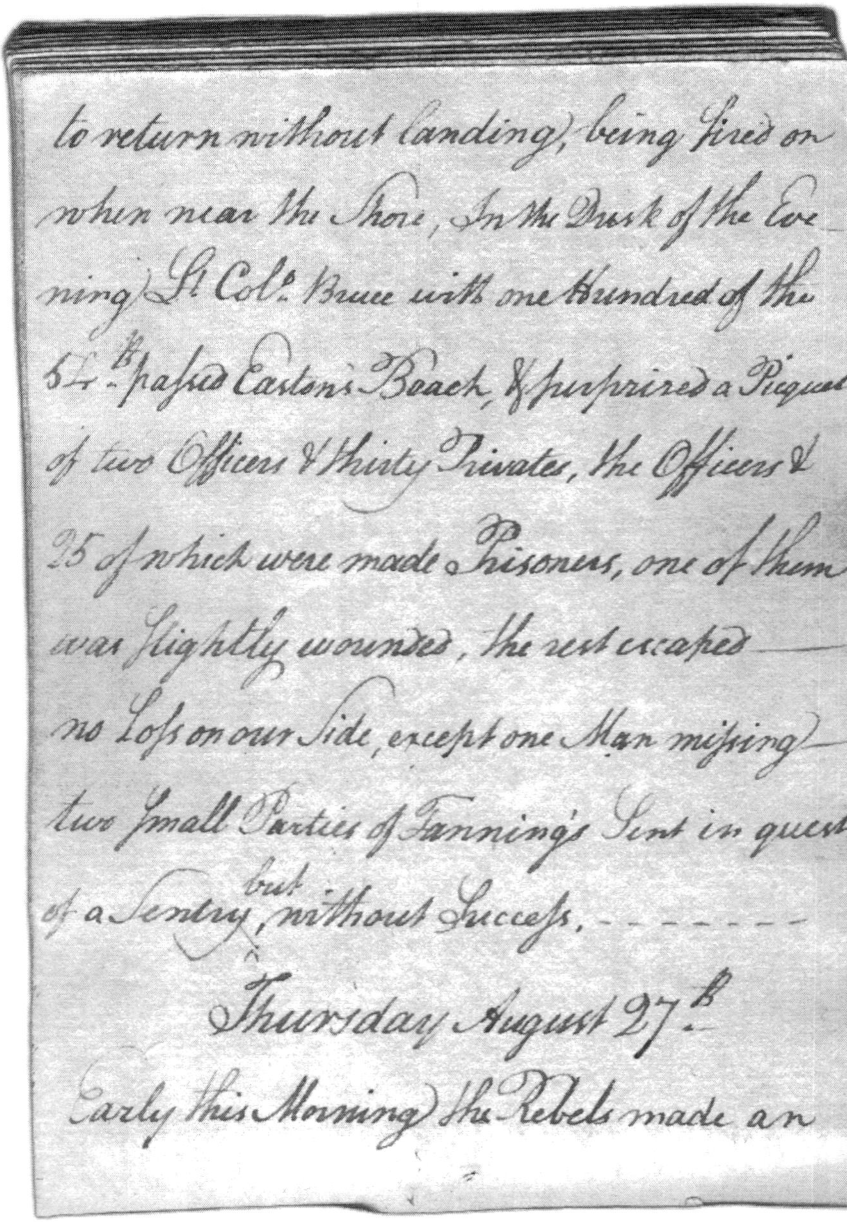
to return without landing, being fired on
when near the Shore, In the Dusk of the Eve-
ning Lt Colo. Bruce with one Hundred of the
54th passed Easton's Beach, & surprized a Picquet
of two Officers & thirty Privates, the Officers &
25 of which were made Prisoners, one of them
was slightly wounded, the rest escaped —
no Loss on our Side, except one Man missing —
two small Parties of Fanning's Sent in quest
of a Sentry, but without Success. - - - - - -

Thursday August 27th.

Early this Morning the Rebels made an

to return without landing, being fired on when near the Shore, In the Dusk of the Evening L^{t}. Col^{o}. Bruce[24] with one Hundred of the 54th passed Easton's Beach, & surprised a Picquet of two Officers & thirty Privates, the Officers & 25 of which were made Prisoners, one of them was slightly wounded, the rest escaped —— no Loss on our Side, except one Man missing — two small Parties of Fanning's sent in quest of a Sentry but without Success -------

Thursday, August 27th

Early this Morning the Rebels made an

24 Andrew Bruce was born in Scotland in 1742, the son of Sir Michael Bruce, who commanded the Stirlingshire Militia in 1745. He was commissioned an Ensign in the new-raising 56th Foot in England (renumbered 54th in 1757). He was promoted to 1st Lieutenant, in the new-raising 85th Foot in England on 9 June 1759, by raising for rank. He became Captain in the 85th Foot in 22 December 1761 by purchase vice a retirement. He went onto the British Half Pay List when his regiment disbanded on 24 May 1763. He transferred from Half Pay as Captain in the 38th Foot in Ireland on 14 February 1765. He was promoted to Major in the 38th Foot on 25 July 1771 by purchase. He became Lieutenant-Colonel in the 54th Foot while serving in Rhode Island on 10 March 1777, by purchase. He was promoted to Brevet Colonel on 20 November 1782. He died on 12 December 1791 at Naples, Italy, after 34.3 years' service.

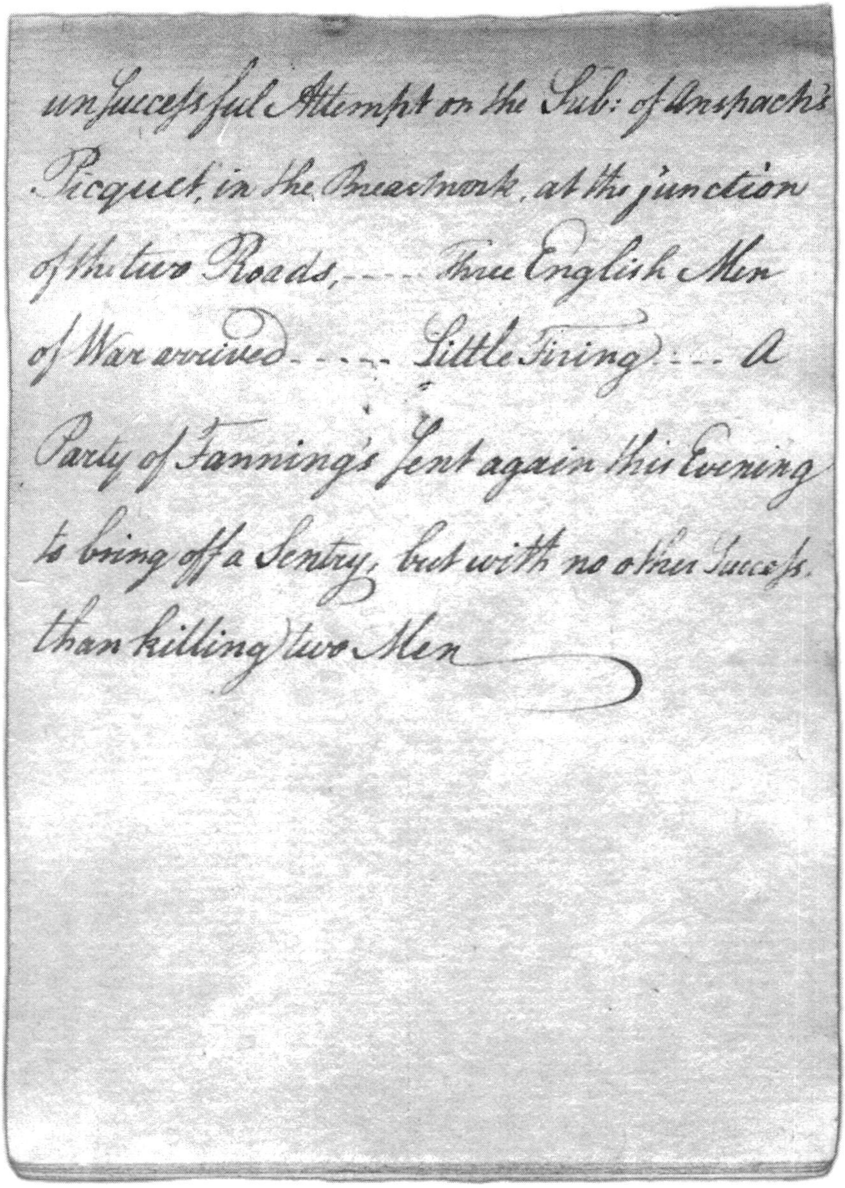
unsuccessful Attempt on the Sub: of Anspach's
Picquet, in the Breastwork, at the junction
of the two Roads,---- Three English Men
of War arrived---- Little Firing---- A
Party of Fanning's sent again this Evening
to bring off a Sentry, but with no other Success
than killing two Men

unsuccessful Attempt on the Sub: of Anspach's Picquet, in the Breastwork, at the junction of the two Roads, ---- Three English Men of War arrived ---- Little Firing ---- A Party of Fanning's sent again this Evening to bring off a Sentry, but with no other Success, than killing two Men.

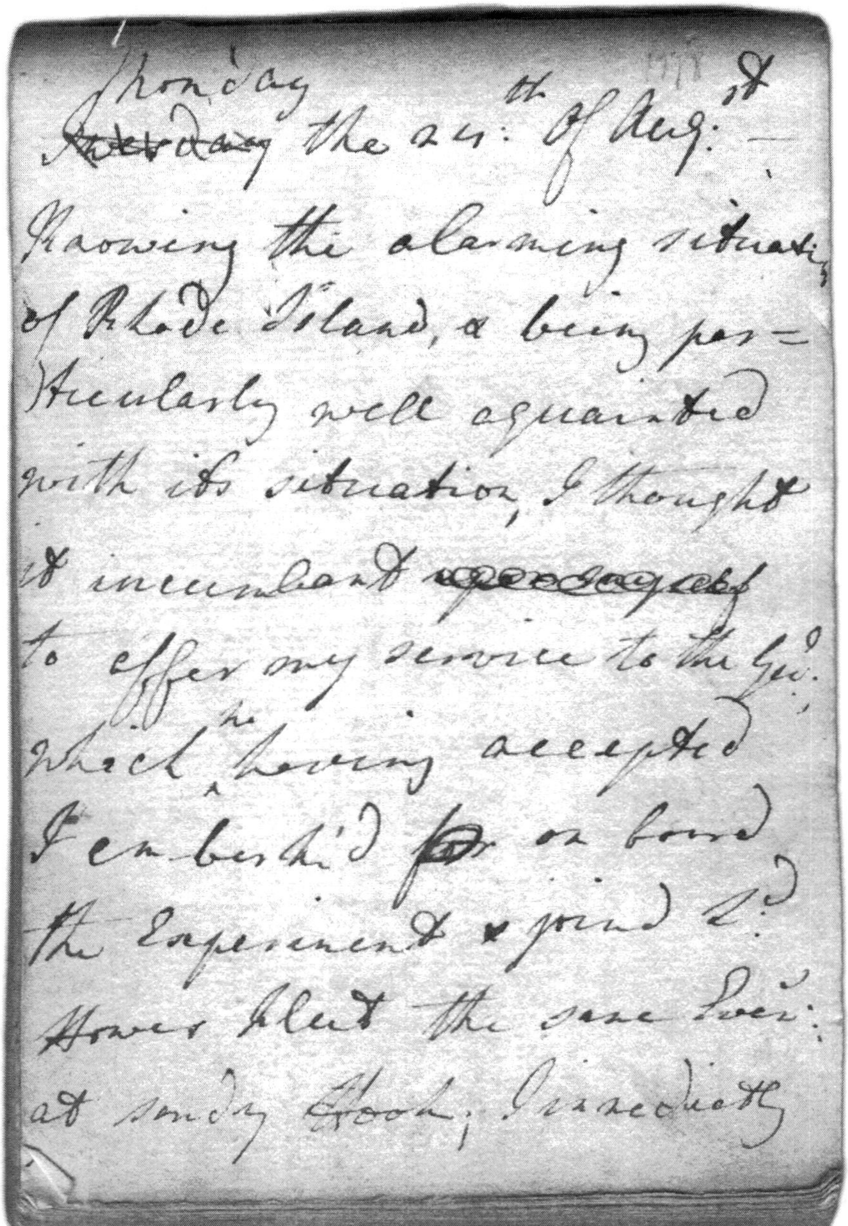

Monday ~~Wednesday~~ the 24th Of Aug: 18

Knowing the alarming situation of Rhode Island, & being particularly well aquainted with its situation, I thought it incumbent ~~[illegible]~~ to offer my service to the Gen.l which he having accepted I embark'd ~~for~~ on board the Experiment & joined Sr. Howes Fleet the same Even: at Sandy Hook; I immediatly

Monday the 24:th of Aug:st

Knowing the alarming situation
of Rhode Island, & being par-
ticularly well aquainted
with its situation, I thought
it incumbent ~~upon myself~~
to offer my service to the Genl:,[25]
which he having accepted
I embark'd ~~for~~ on board
the Experiment[26] & joind L^{d}.
Howes fleet the same even^{g}
at Sandy Hook; I immediately

---

25 General Sir Henry Clinton

26 The 50-gun HMS *Experiment* was commanded by Captain Sir James Wallace from 4 June 1776 to 24 September 1779. He is more widely remembered in Rhode Island for his previous command, the 20-gun frigate *Rose*, from 15 November 1771 to 4 June 1776. On 20 August 1778, *Experiment* was on detached duty off Newport when d'Estaing's fleet arrived. Escaping to the West, Wallace sailed down Long Island Sound and, in a feat of seamanship not previously accomplished by such a large vessel, he took his ship through Hell Gate and into the East River to reach New York City. Built in 1772–1774 at Deptford, the ship was commissioned in 1775, Her overall length was 140 feet 9 inches overall, length of keel 115 feet 6 inches, breadth 38 feet 9 inches, draft 16 feet 7 inches, 910 60/94 BM.

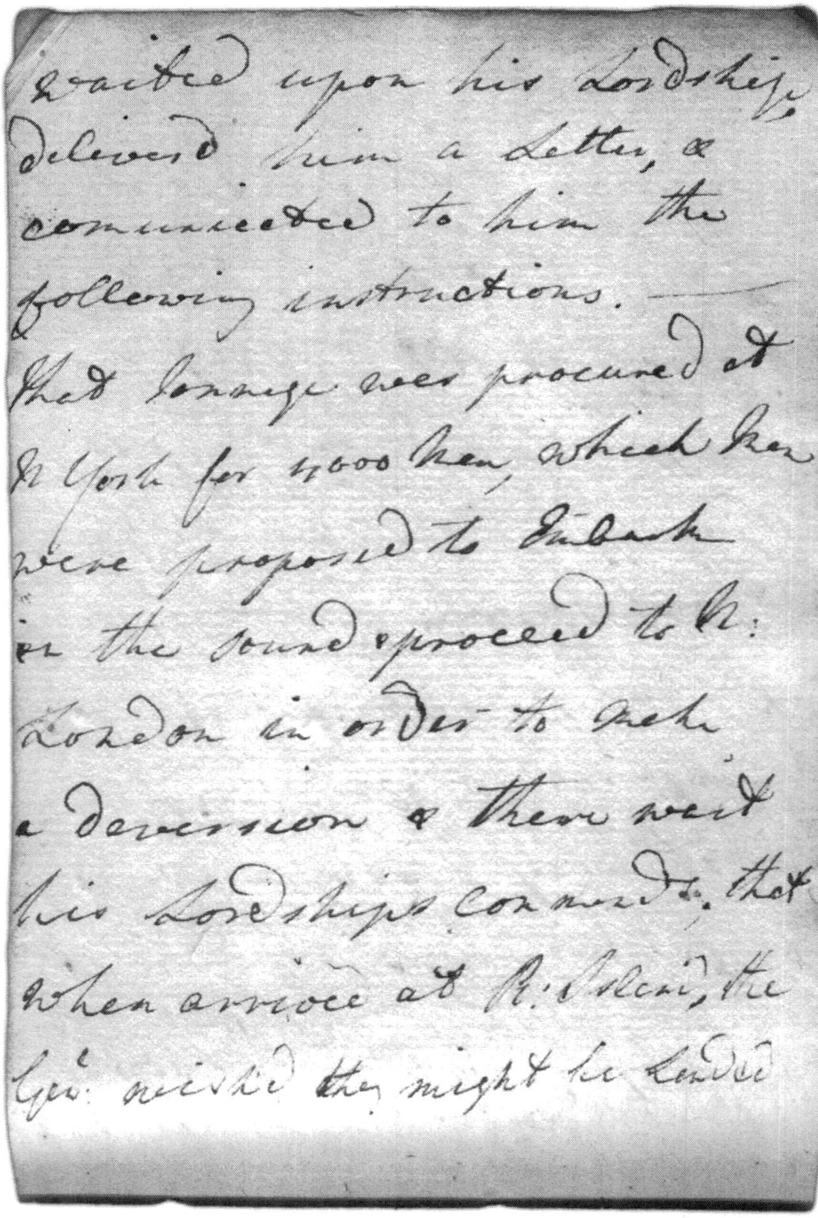

waited upon his Lordship, delivered him a Letter, & communicated to him the following instructions. —
That tonnage was procured at N York for 4000 men, which men were proposed to Embark in the sound & proceed to N: London in order to make a diversion & there wait his Lordships commands, that when arrived at R: Island, the Gen: wished they might be Landed

waited upon his Lordship,
deliverd him a letter, &
comunicated to him the
following instructions. —
That tonnage was procured at
N York for 4000 Men, which Men
were proposed to Embark
in the sound & proceed to N.
London in order to make
a deversion & there wait
his Lordships commands, that
when arrived at R. Island, the
Genl: wished they might be Landed

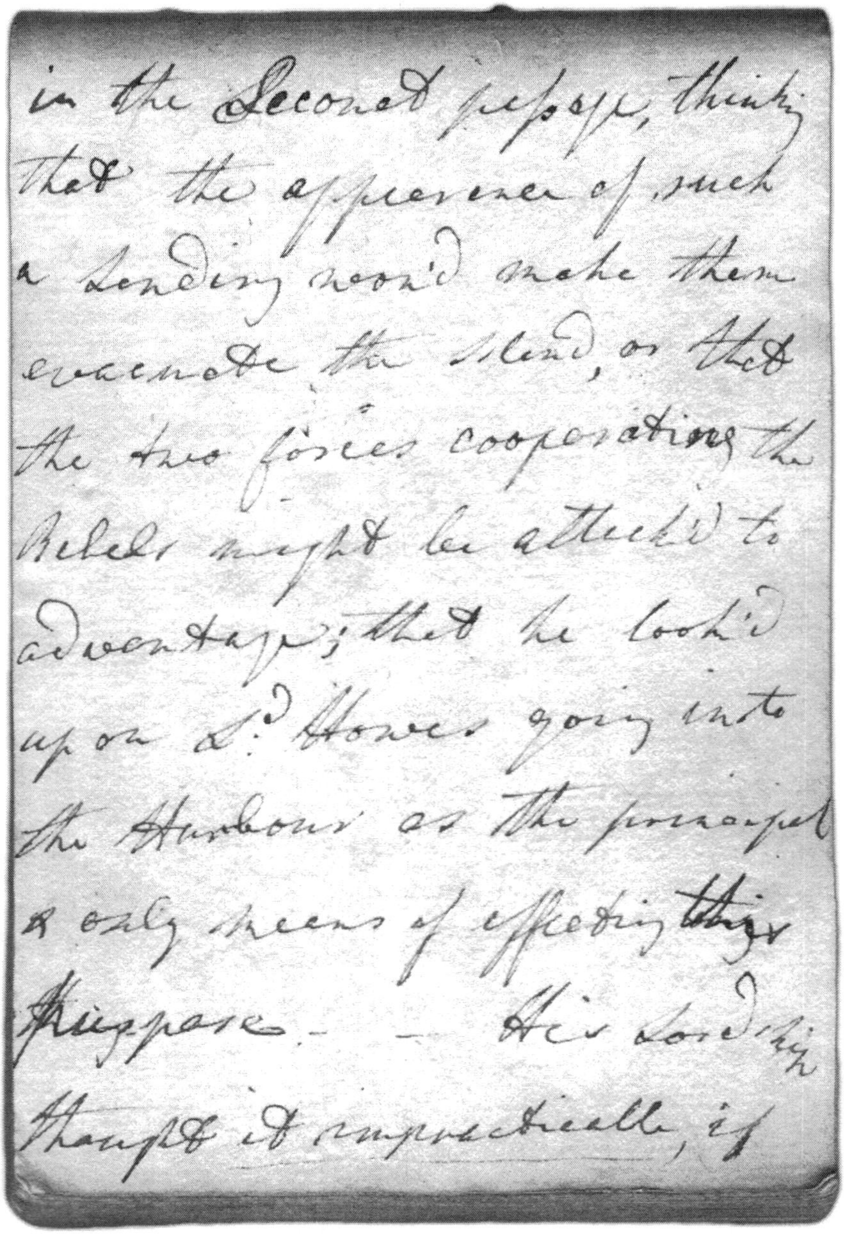

in the Second purpose, thinking
that the appearance of such
a Landing would make them
evacuate the Island, or that
the two forces cooperating the
Rebels might be attack'd to
advantage; that he look'd
upon Ld Howes going into
the Harbour as the principal
& only means of effecting this
purpose. — His Lordship
thought it impracticable, if

in the Seconet passage, thinking
that the appeerence of such
a Landing wou'd make them
evacuate the Island, or that
the two forces cooperating the
Rebels might be attack'd to
advantage; that he look'd
up on L:[d] Howes going into
the Harbour as the principal
& only means of effecting this
purpose. – – His Lordship
thought it impracticable, if

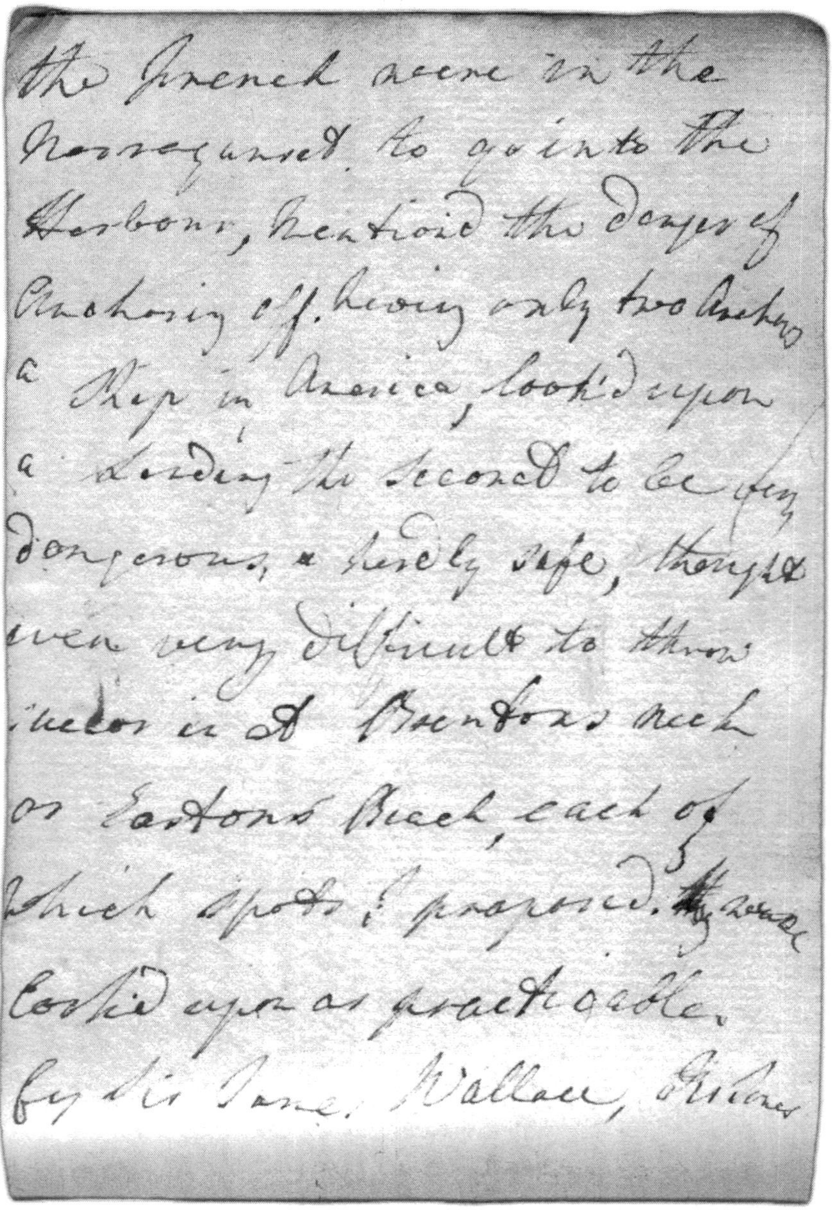

the French were in the
Narraganset to go into the
Harbour, mention'd the danger of
Anchoring off, having only two Anchors
a Ship in America, look'd upon
a Landing the second to be very
dangerous, & hardly safe, though
even very difficult to throw
succor in at Brentons neck
or Eastons Beach, each of
which spots I proposed, [illegible]
look'd upon as practicable.
by Sir Jame. Wallace, Fisher

the French were in the
Narraganset to go in to the
Harbour, mention'd the danger of
Anchoring off having only two Anchors
a Ship in America, look'd upon
a Landing the Seconet to be very
dangerous, & hardly safe, thought
even very difficult to throw
anchor in at Brenton's Neck
or Easton's Beach, each of
which spots I proposed that were
look'd upon as practicable by Sir James Wallace,[27] [illegible]

27 Captain Sir James Wallace commanding HMS *Experiment.*

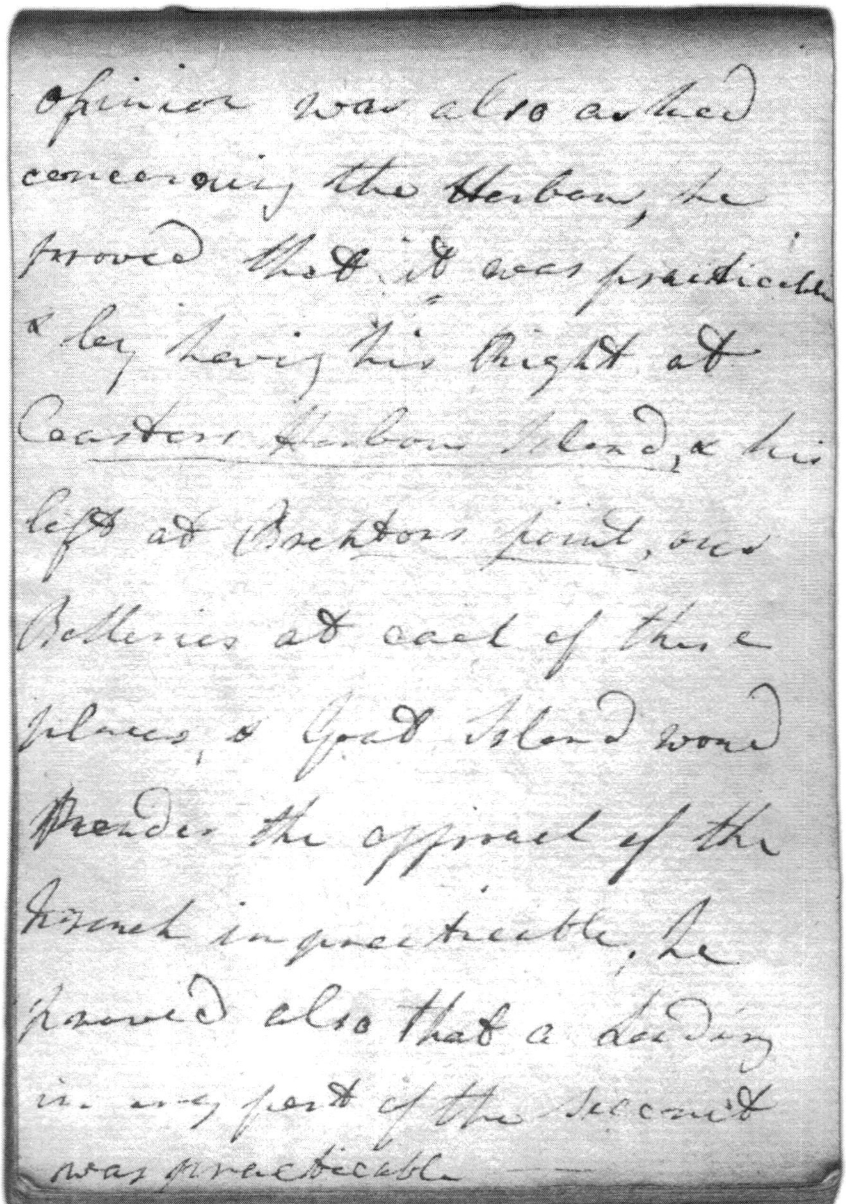

opinion was also asked
concerning the Harbour, he
proved that it was practicable
& by having his Right at
Coasters Harbour Island, & his
left at Brentons point, our
Batteries at each of these
places, & Goat Island would
render the approach of the
French impracticable, he
proved also that a Landing
in any part of the Continent
was practicable ——

opinion was also asked concerning the Harbour, he proved that it was practicable & by having his Right at Coasters Harbour Island, & his left at Brenton's point, our Batteries at each of these places, & Goat Island would render the approach of the French impracticable; he proved also that a Landing in any part of the Seaconet was practicable. ---

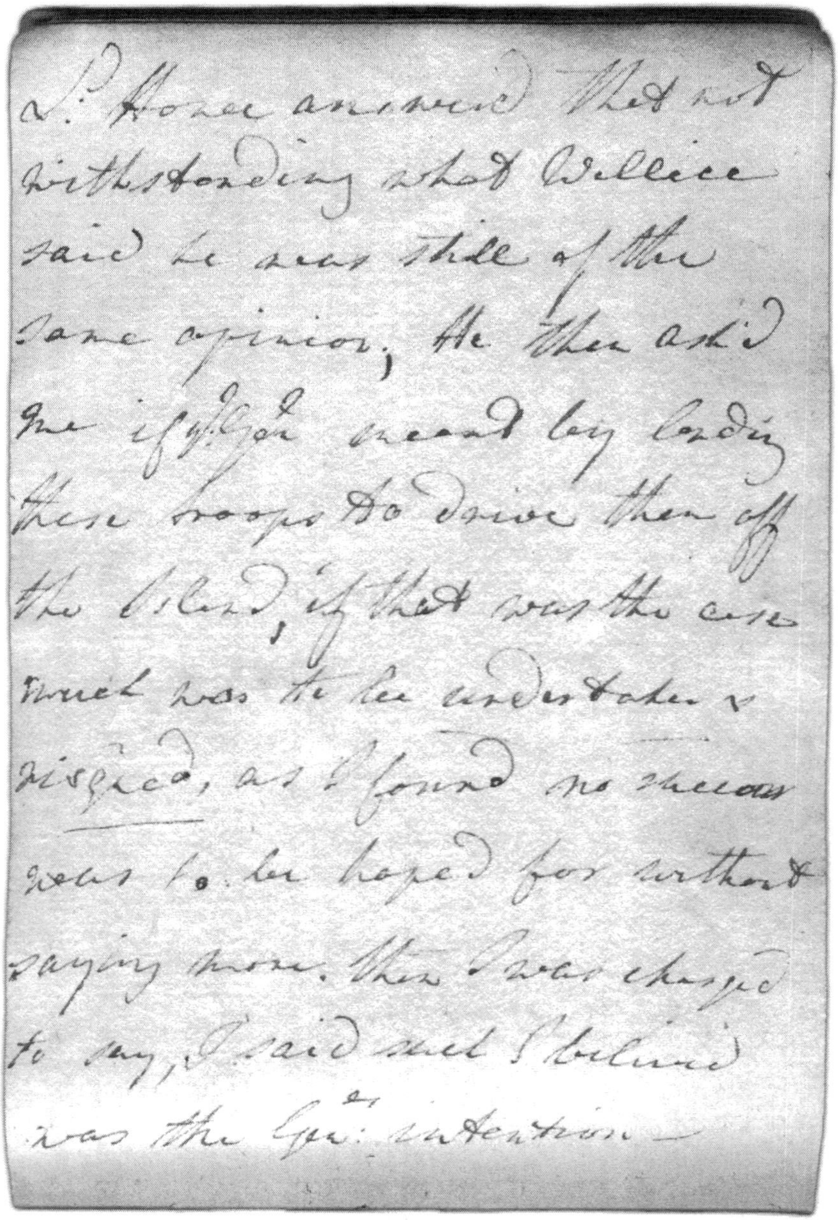

& Sr. Howe answered that not
withstanding what Wallace
said he was still of the
same opinion, He then ask'd
me if Sr. Wm. meant by landing
these troops to drive them off
the Island, if that was the case
which was to be undertaken &
risqued, as I found no success
was to be hoped for without
saying more, then I was charged
to say, I said such I believed
was the Genr. intention —

L:[d] Howe answer'd that notwithstanding what Wallace said he was still of the same opinion; He then ask'd me if ye Gen[l]: meant by landing these troops to drive them off the Island, if that was the case much was to be undertaken & risqued, as I found no meens was to be hoped for without saying more then I was charged to say, I said such I believed was the Gen[ls]: intention. —

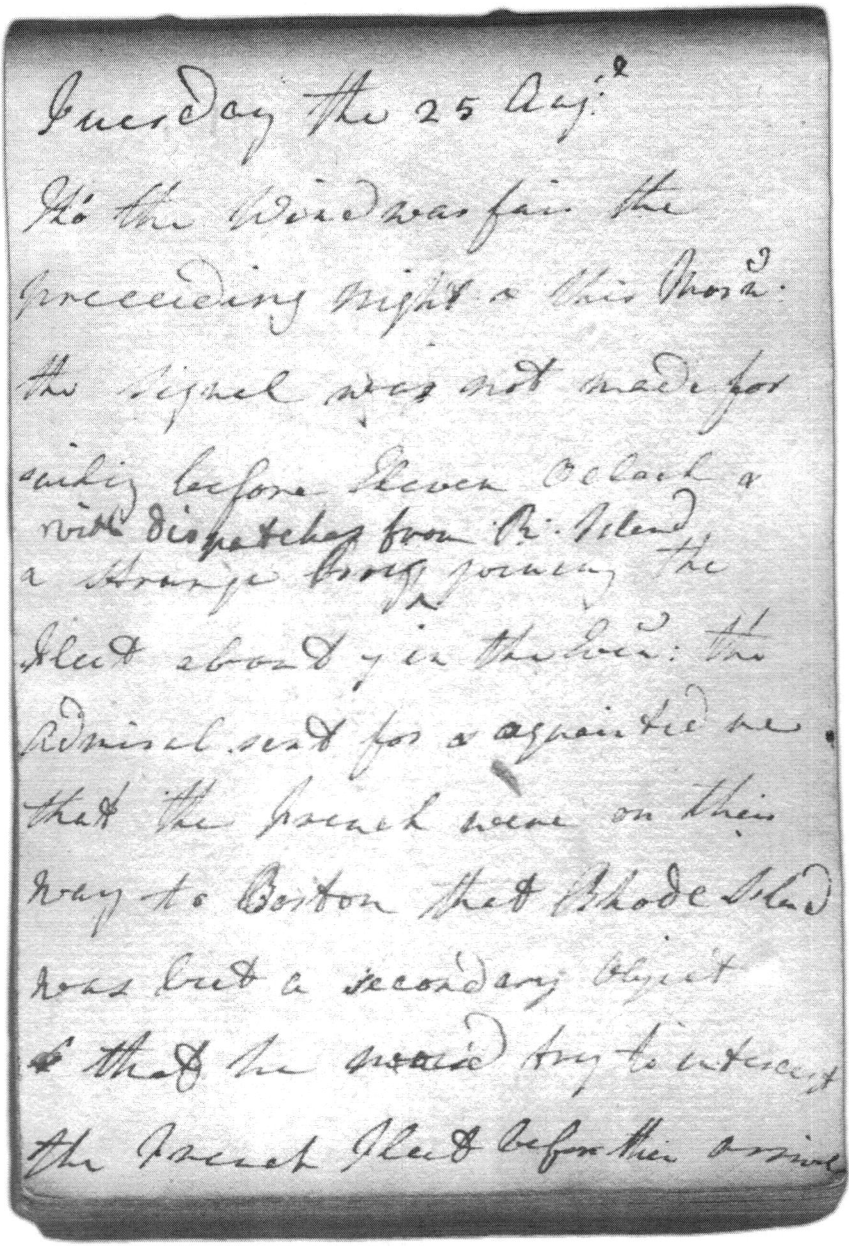
Tuesday the 25 Aug.st
Tho' the Wind was fair the
preceeding night & this morn.g
the Signal was not made for
sailing before Eleven O'clock &
with dispatches from R: Island
a Strange brig joining the
Fleet aboard & in the River: the
Admiral sent for & acquainted me
that the French were on their
way to Boston that Rhode Island
was but a secondary object
& that he would try to intercept
the French Fleet before their arrival

Tuesday the 25 Aug.st

Tho' the Wind was fair the
preceeding night & this Morng
the signel was not made for
sailing before Eleven o'clock &
a strange brig with dispatches from R. Island joining the
fleet about 7 in the Eveng. the
Admiral sent for & aquainted me
that the French were on their
way to Boston that Rhode Island
was but a secondary object
& that he wou'd try to intercept
the French Fleet before their arrival,

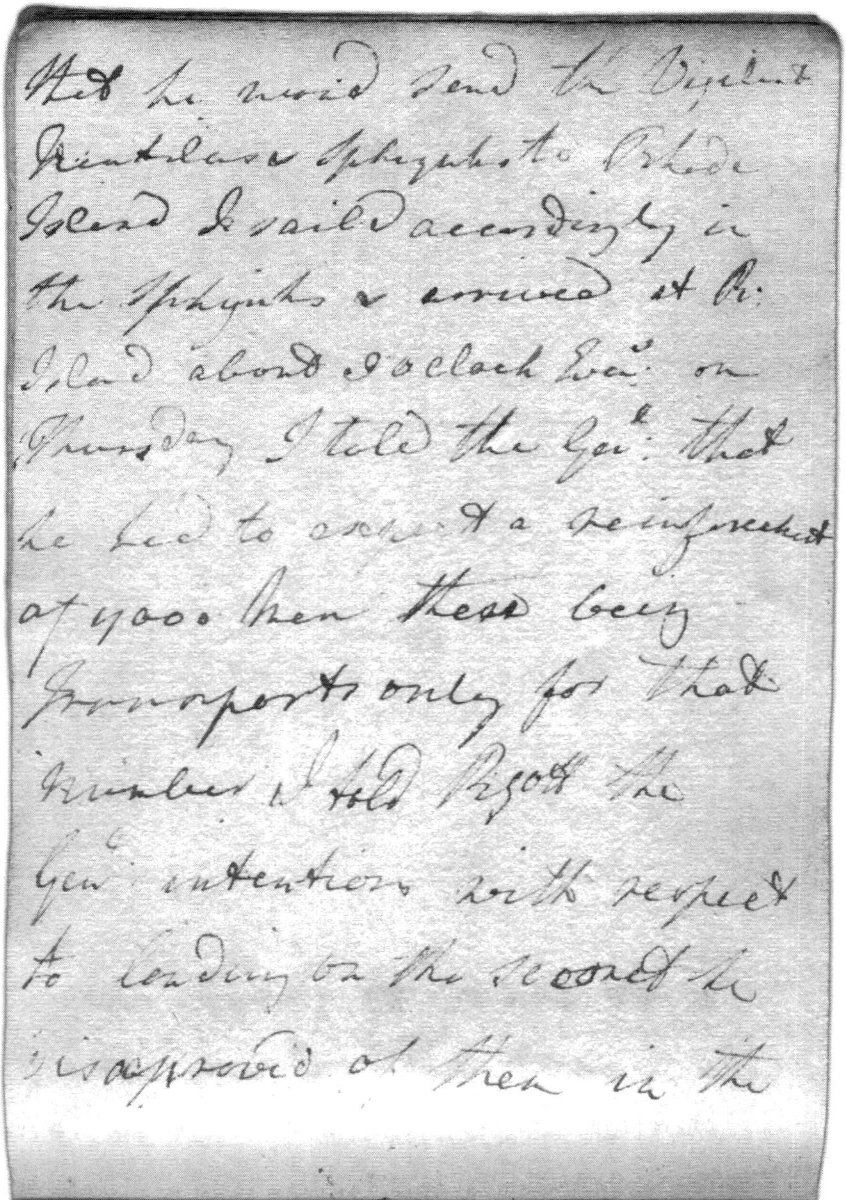
that he would send the Diligent
[illegible] & Sphynx to Rhode
Island I sail'd accordingly in
the Sphynx & arrived at Rh:
Island about 8 o'clock Even: on
Thursday I told the Gen: that
he had to expect a reinforcement
of 4000 men these being
transports only for that
number I told Pigott the
Gen: intentions with respect
to landing on the Island he
disapproved of them in the

that he would send the Vigilant,[28]
Nautilas,[29] & Sphynks[30] to Rhode
Island. I saild accordingly in
the Sphynks & arrived at R.
Island about 9 o'clock Even^{g}. on
Thursday. I told the Genl: that
he had to expect a reinforcement
of 4000 men there being
transports only for that
number. I told Pigott[31] the
Genl: intentions with respect
to landing on the Seconet he
disaproved of them in the

28 The 22-gun *Vigilant* was the former merchant ship *Empress of Russia* that the Admiralty had purchased in 1777 and designated a Sixth Rate. She was under the command of Commander Brabazon Christian.

29 The 16-gun sloop *Nautilus* was commanded by Commander John Becher.

30 The 20-gun frigate *Sphinx* was the lead ship in the Sphinx Class. She was commanded by Captain Alexander Graeme.

31 See note in Addendum on p. 139.

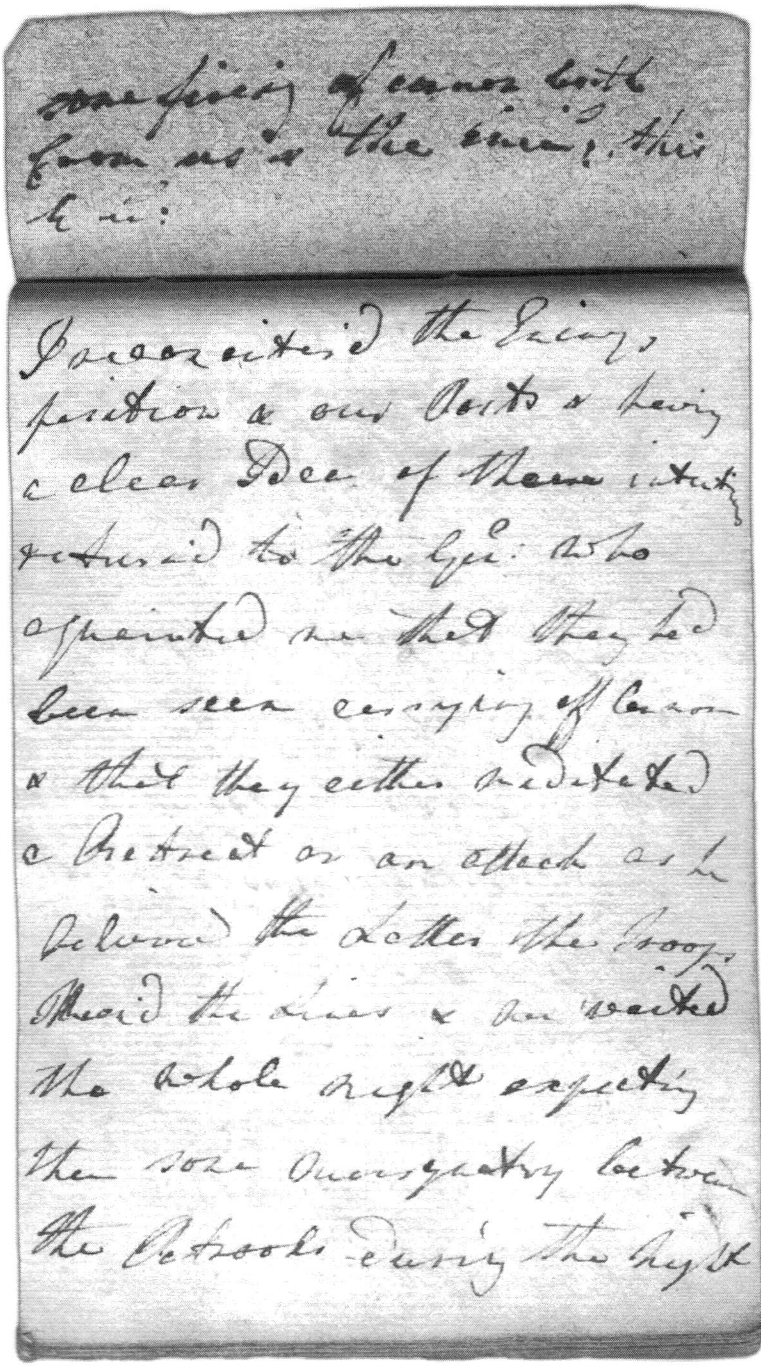

some firing of cannon both from us & the enemy, this [illegible]:

I reconnoitred the Enemys position & our Posts & having a clear Idea of their situation returned to the Gen: who acquainted me that they had been seen carrying off Cannon & that they either meditated a Retreat or an attack as he believed the Latter the Troops Manned the Lines & we waited the whole night expecting them some Musquetry between the Patrools during the night

some firing of canon both
from us & the Enimy this
Even^{g}.
I reconoiter'd the Enimy's
position & our Posts & having
a cleer Idea of *these* intentions
&c huried to the Genl: who
aquainted me that they had
been seen carrying off Canon
& that they either meditated
a Retreat or an attack as he
believed the Latter the troops
[illegible] the lines & we waited
the whole night expecting
then some musquetry between
the Outworks during the Night.

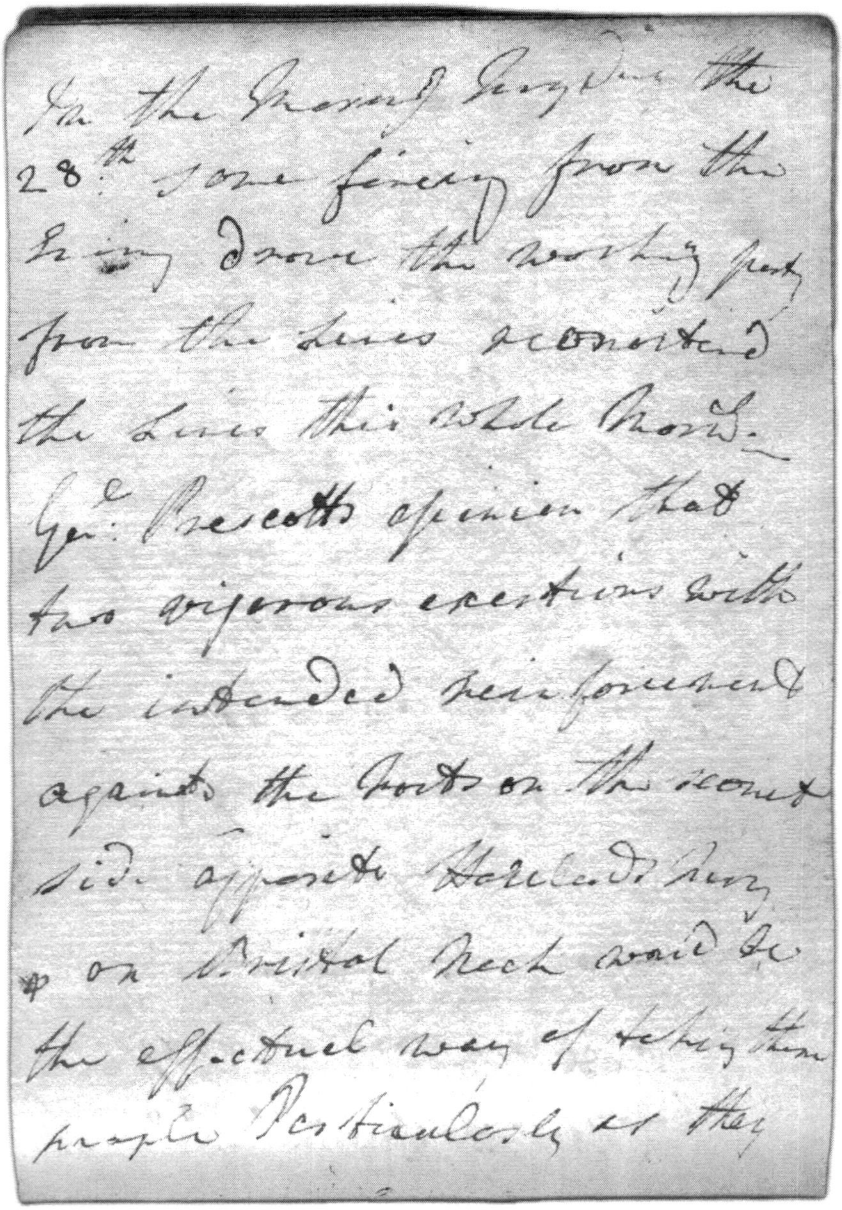

In the Morning of Friday the
28th some firing from the
Enemy drove the working party
from the Lines reconnoitred
the Lines their whole length.
Genl. Prescotts opinion that
two vigorous exertions with
the intended reinforcement
against the Forts on the Island
side opposite Howlands ferry
& on Bristol Neck would be
the effectual way of taking them
people Particularly as they

In the Morn'g Fryday the
28:[th] some fireing from the
Enimy drove the working party
from the Lines reconoiter'd
the Lines this whole Morn[g].
Gen[l]: Prescotts opinion that
two vigorous exertions with
the intended reinforcement
against the Forts on the Seconet
side opposite Howland's Ferry
& on Bristol Neck wou'd be
the effectual way of taking there
people Particularly as they

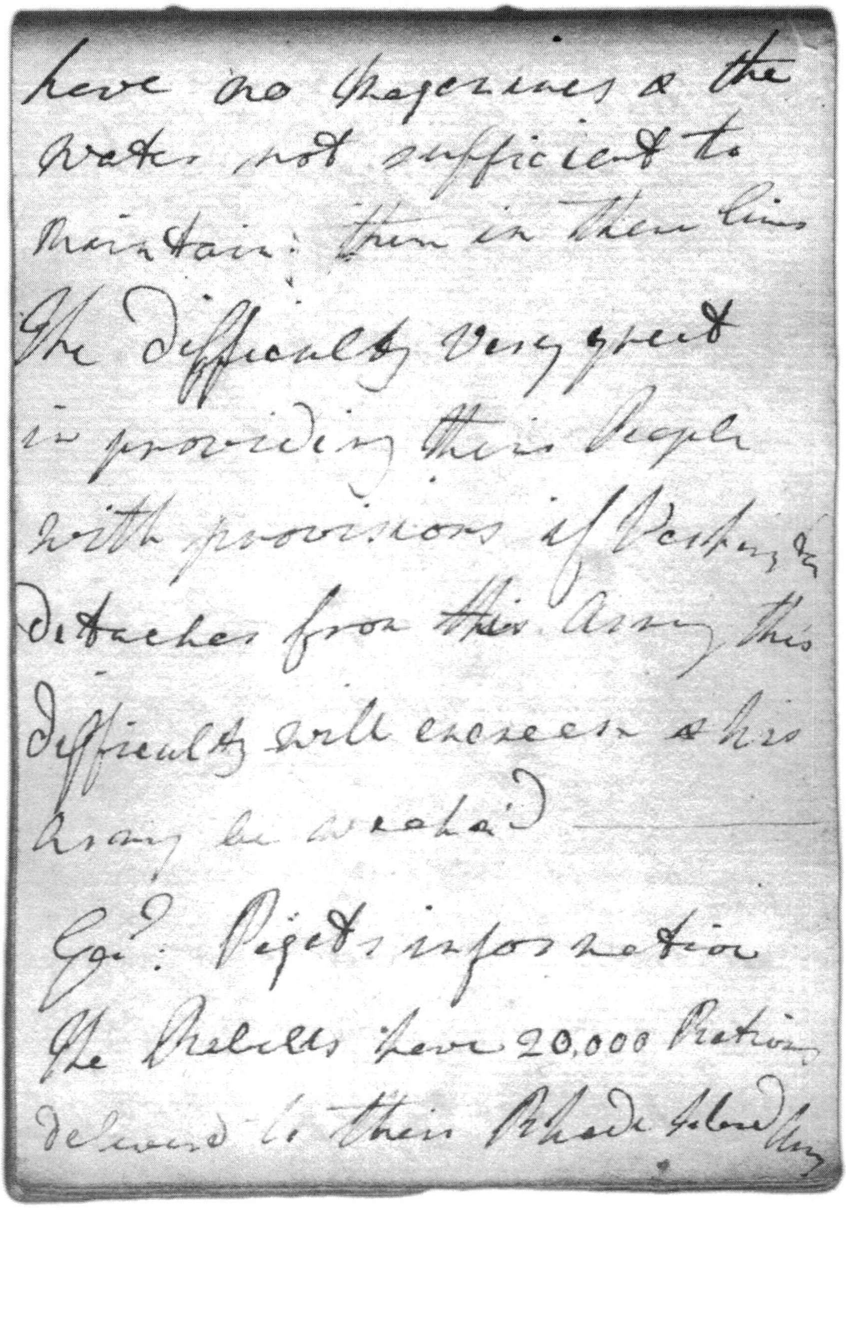

have no Magazines & the
water not sufficient to
maintain them in their lines
The difficulty very great
in providing their People
with provisions if Parties are
detached from their Army this
difficulty will encrease & his
Army be weaken'd ——
Genl: Pigots information
The Rebells have 20,000 Rations
delivered to them Rhode Island Army

have no megezines & the
Water not sufficient to
maintain them in their lines.
The difficulty very great
in providing their People
with provisions. if Washington
detaches from this army this
difficulty will encreese & his
Army be weakn'd ——
Gen[l]: Pigots information
The Rebells have 20,000 Rations
delivered to their Rhode Island Army.

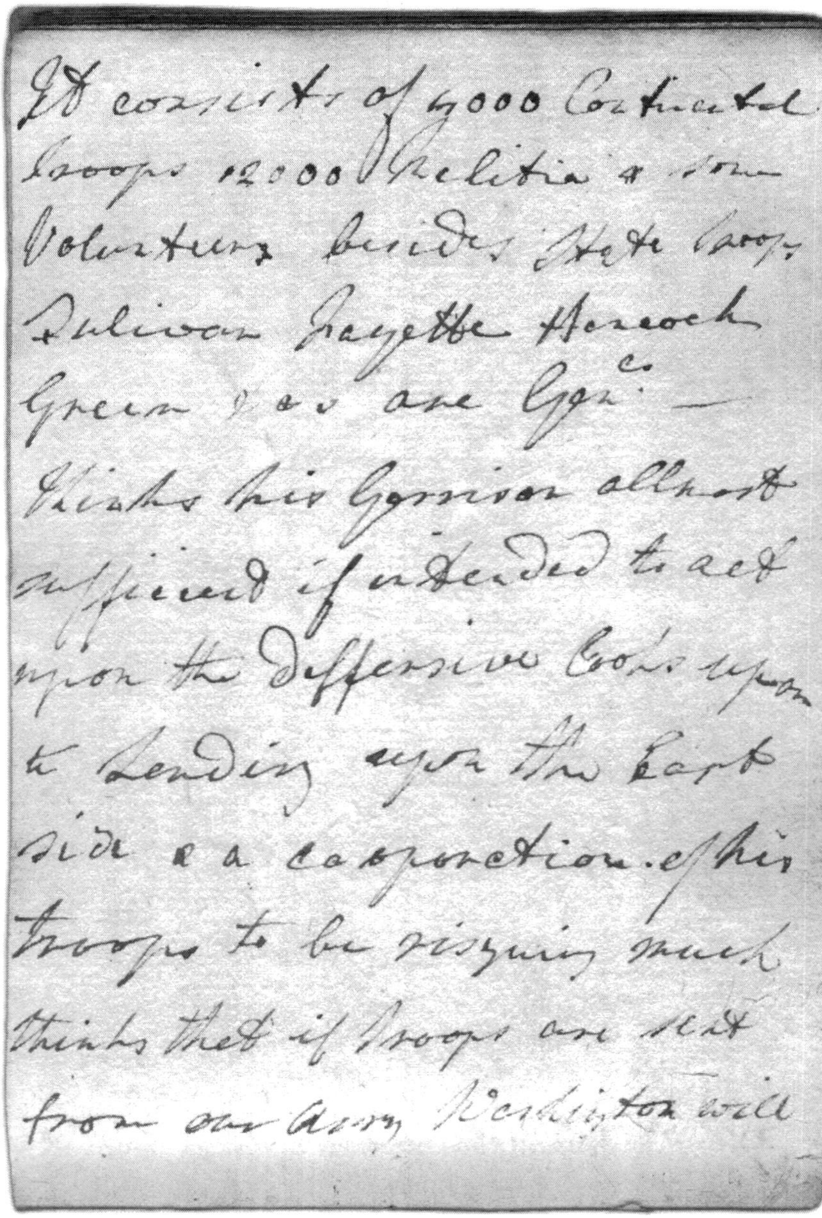

It consists of 4000 Continental
troops 12000 Militia & some
Volunteers besides State troops
Sullivan Fayette Hancock
Greene &c are Genls —
thinks his Garrison allmost
sufficient if intended to act
upon the deffensive Looks upon
a Landing upon the East
side & a cooperation of his
troops to be requiring much
thinks that if troops are sent
from our Army Washington will

It consists of 4000 Continental
troops 12000 Melitia & some
Volunteers besides State troops.
Sulivan, Fayette, Hancock,
Green, etc. are Gen.[ls]. —
thinks his Garrison allmost
sufficient if intended to act
upon the deffensive. looks upon
a Landing upon the East
side & a *co oporetion* of his
troops to be risquing much.
thinks that if troops are sent
from our Army Washington will

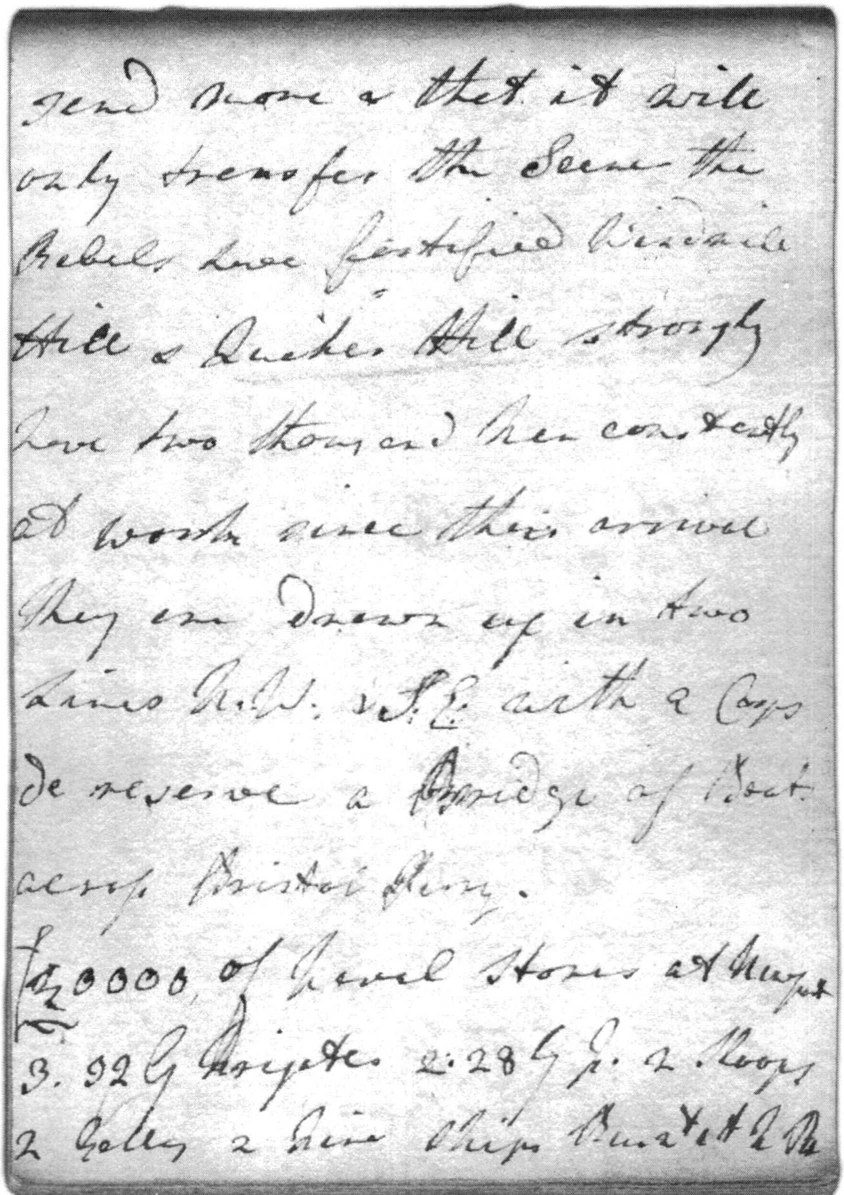
send more & that it will
only transfer the Scene the
Rebels have fortified [illegible]
Hill & Quaker Hill strongly
have two thousand men constantly
at work since their arrival
they are drawn up in two
Lines N.W. & S.E. with a Corps
de reserve a Bridge of Boats
across Bristol Ferry.
£20000 of Naval Stores at Newport
3. 32 G frigates 2: 28 G f. 2 Sloops
2 Galleys 2 Fire Ships Burnt at R. I.

send more & thet it will
only transfer the Scene the
Rebels have fortified Windmill
Hill & Quaker Hill strongly
have two thousand Men constently
at work since their arrival.
They are drawn up in two
Lines N.W. & S.E. with a Corps
de reserve a Bridge of Boats
across Bristol Ferry.

£40000 of Naval stores at Newport,
3.32 G. Frigates, 2.28 G.F., 2 Sloops,
2 Gally 2 Fire ships burnt at N Pt.

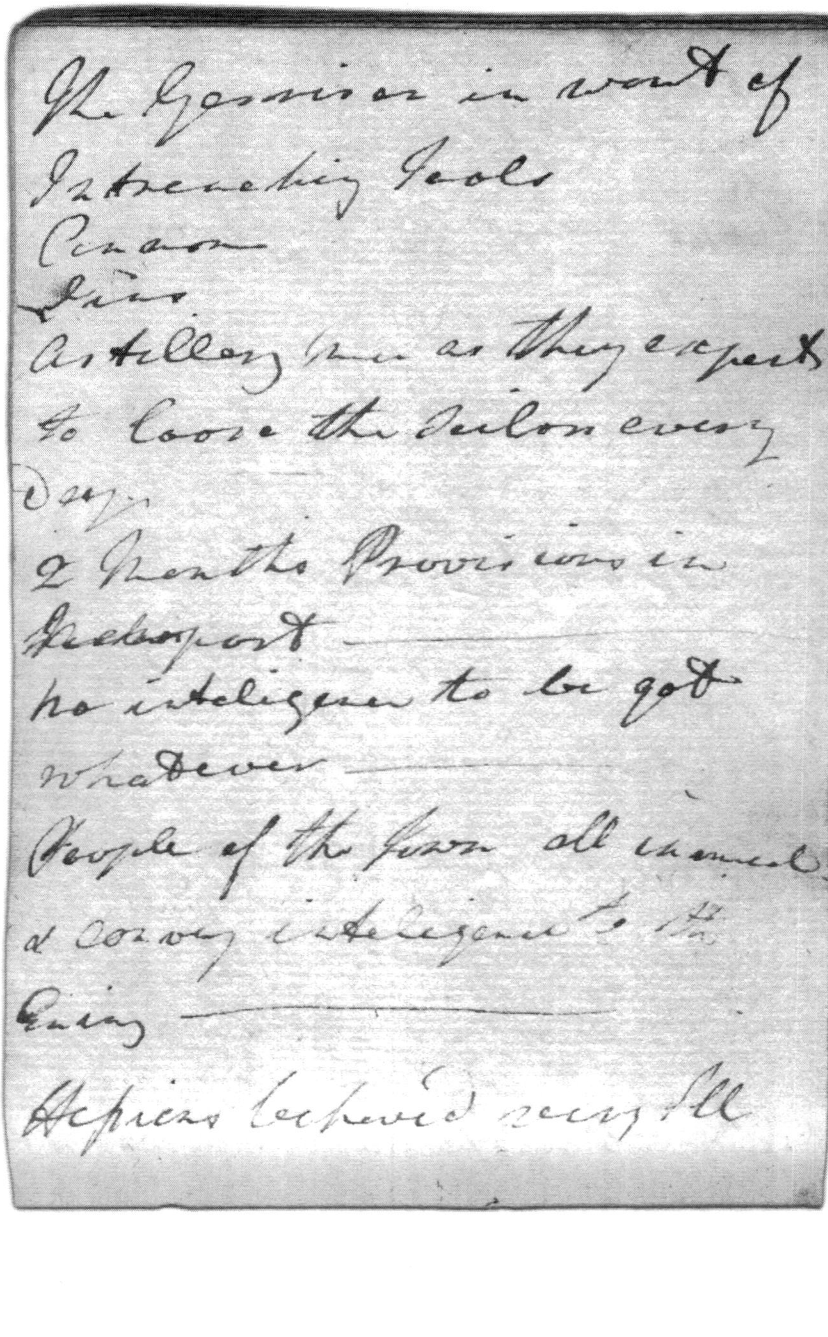
The Garrison in want of
Intrenching Tools
Cannon
Fire
Artillery men as they expect
to loose the Sailors every
Day
2 Months Provisions in
[illegible]port ———
no inteligence to be got
whatever ———
People of the Town all inclined
& convey inteligence to the
Enemy ———
Hessians behaved very ill

The Garrison in want of
Intrenching Tools
Canon
[? lines]
Artillery men as they expect
to loose the sailors every
day.
2 Months Provisions in
Newport ——————
No inteligence to be got
Whatever ——————
People of the Town all *[unarmed?]*
& convey inteligence to the
enimy. ——————
Hessiens behaved very *[ill] III*

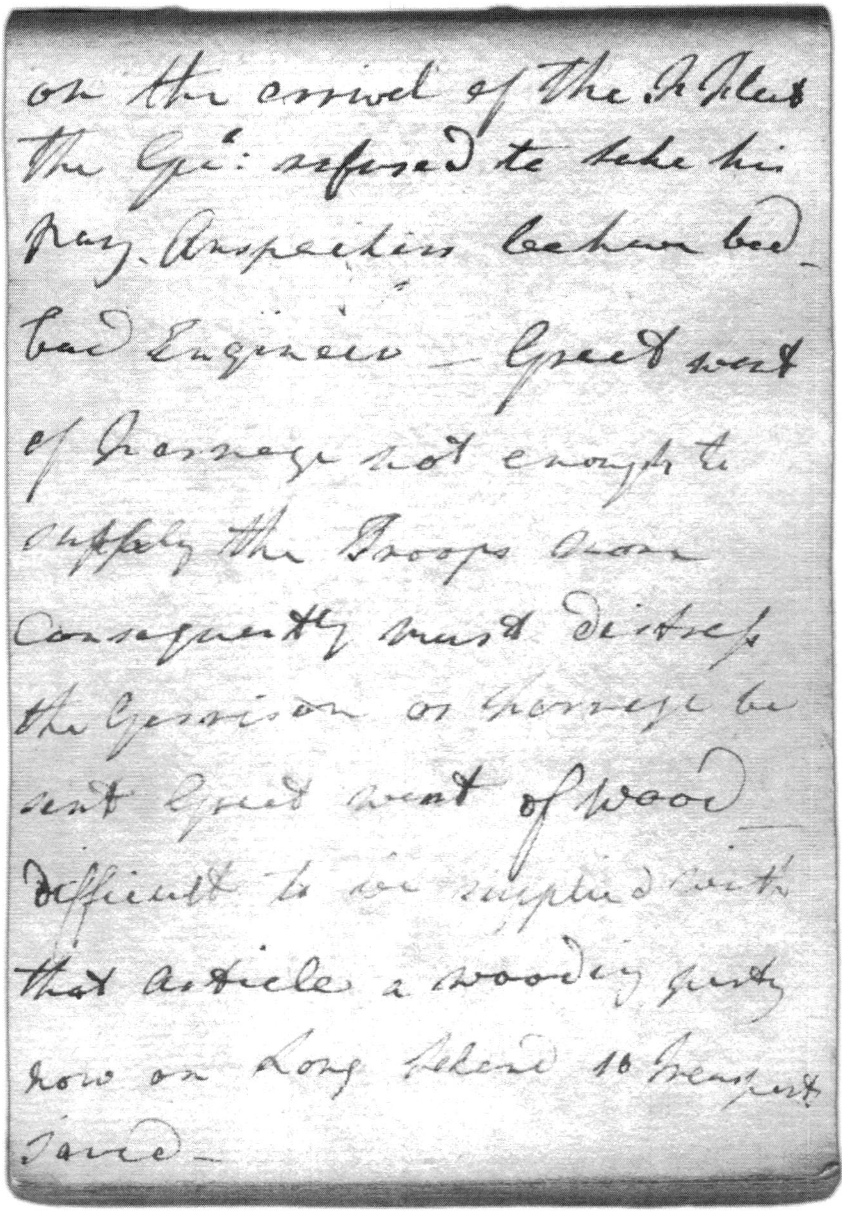

on the arrival of the 2d Fleet
the Genl: refused to take his
Majs. Inspection between bad-
bad Engineers — Great want
of Forrage not enough to
supply the Troops now
Consequently must distress
the Garrison or Forrage be
sent Great want of Wood
difficult to be supplied with
that Article a wooding party
now on Long Island 10 Transports
saved —

on the arrival of the F. Fleet.
the Gen[l]: refused to take his
pay. Anspechers behave bad –
bad Engineer — Great want
of Forrage not enough to
supply the Troops now
Consequently must distress
the Garrison or forrage be
sent. Great want of wood —
difficult to be supplied with
that article. a wooding party
now on Long Island. 10 transports
saved. –

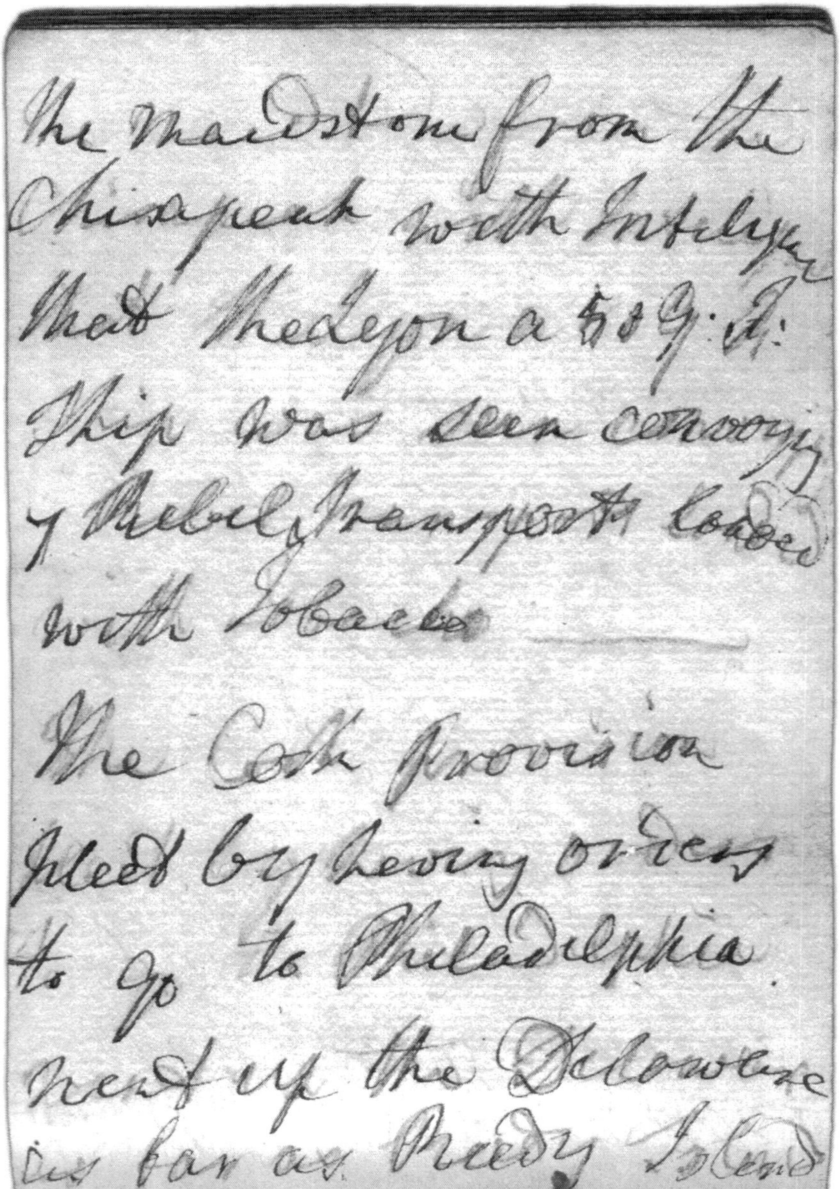
the Maidstone from the
Chisapeak with Intellige[nce]
that the Lyon a 50 G: S:
Ship was seen convoying
y Rebel Transports loaded
with Tobacco

The Cork provision
fleet by having orders
to go to Philadelphia
went up the Delaware
as far as Reedy Island

The MaidStone[32] from the
Chisapeak with Intelgence
that the Lejon a 50 G:F:[33]
Ship was seen convoying
7 Rebel Transports loaded
with Tobacco ————

The Cost provision
Fleet by having orders
to go to Philadelphia
went up the Delaware
as far as Reedy Island

32 The 36-gun frigate *Maidstone* was under the command of Captain Alan Gardner.

33 Not identified. No French warship of this or a similar name has been identified as being in American waters at this time.

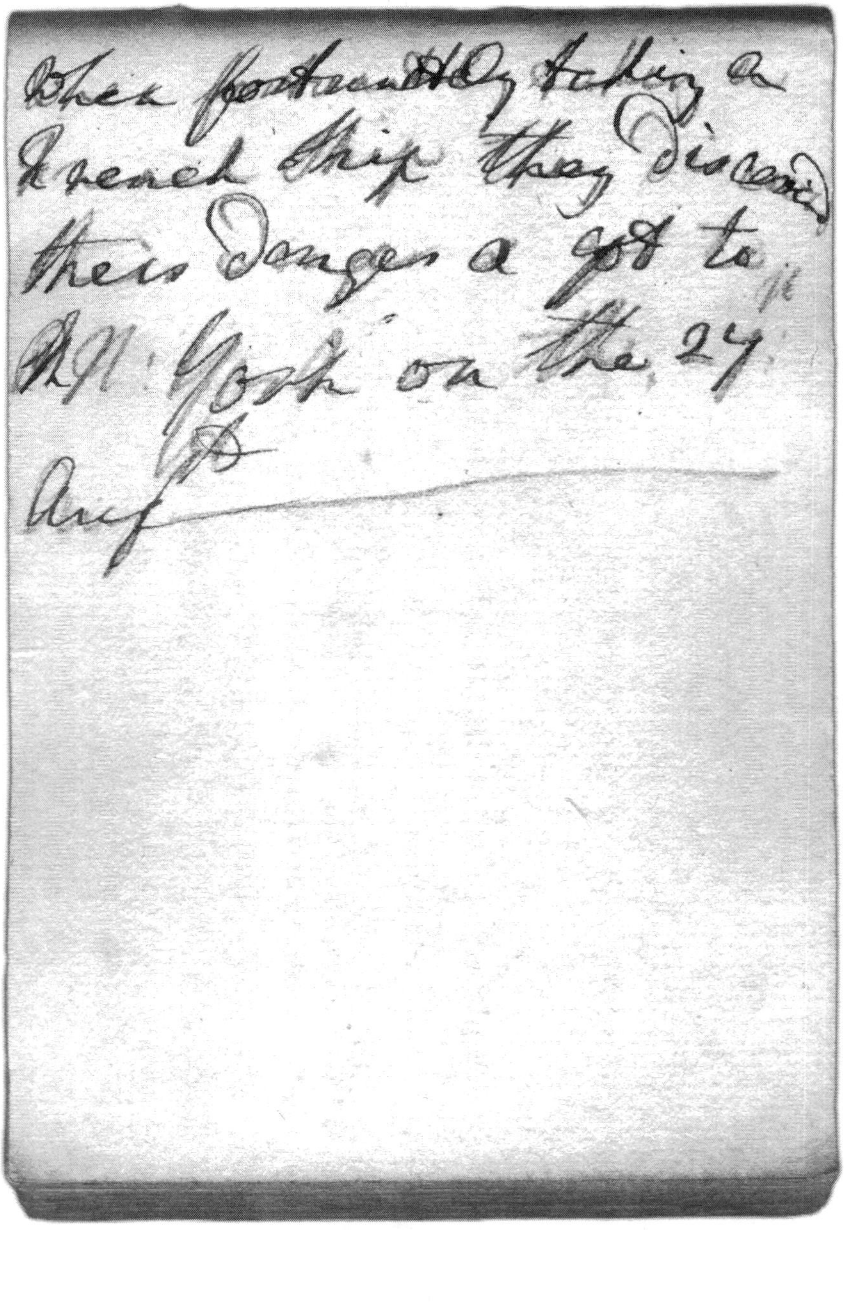

When fortunately taking a
French ship they discovered
their danger & got to
N. York on the 27
Aug

when fortunately taking a
French Ship they discern'd
their danger & got to
N York on the 27th
Augst. ————

Mrs Jameson

Mrs Marshon[34]

34 Not identified.

PART TWO

Letters and Other Documents

Letters and Other Documents

Lieutenant-Colonel Charles Stuart wrote the following letters either at about the same time that he was keeping the diary reproduced in Part I, or shortly afterward recording more detail or other aspects concerning the operations in Rhode Island. The original copies of the letters from which these transcriptions were made in the 1920s have not been found in any publicly available collection. They may possibly still exist in an unknown private collection or they may be entirely lost. The following transcriptions are taken from the uncopyrighted, anonymously edited, New Records of the American Revolution, *privately printed in 1927. All the letters in that volume were from the Charles Stuart family papers that were at Highcliffe Castle*

1. Lieutenant-Colonel Stuart to Lord Bute, 22 August 1778.[1]

1 Anonymous, ed., *New Records of the American Revolution: the Letters, Manuscripts and Documents sent by Sir Charles Stuart to the Earl of Bute, also the letters of General Howe, General Clinton, and Other Officers to Sir Charles Stuart* (London: L. Kashner, 1927), p. 54.

Our affairs, Dearest Father, look more & more dismal; a Man of War[2] which was sent by Lord Howe for the relief of Rhode Island is return'd or rather drove by a French line of battle ship up the Sound & the following unexpected account gives all reason to conjecture that that Island either is or in a very few days will be in possession of our Enemies.

On Thursday last Sir James Wallace who commanded the ship sent out from our Fleet perceived thirteen French ships sailing into Rhode Island, one appeared to be tow'd & considerably shattered; accounts were also brought that the Rebels had some days before landed at Howlands Ferry 19,000 Men, that they had actually begun regular approaches against our lines & were only waiting this return of the French to bring their scheme to a conclusion.

Notwithstanding many frigates were cruising in watch of the French, Ld. How was so unfortunate as not to get immediate intelligence of their movement from the Delaware, and I am very much afraid that, seeing Lord Howe and the French Fleet in pursuit fo them and after a week finding the French returned without molestation, that the Garrison will naturally infer from these appearances the defeat of the British Squadron, and I am convinced that even a certainty of it wou'd not discourage the British, yet as there are but 2,000 of them, the rest being Hessians, Anspach's, Islanders and Seamen (most of whom are too apt to be disheartened) I must conclude it my own opinion that if the Rebels have acted with their usual industry Rhode Island must fall before our Fleet

2 HMS *Experiment*

can arrive. Unfortunately only two of Byron's Fleet are arrived. The account of Pigot's situation and the F. Fleet arrived 3 days ago, since which time the winds have been hostile and Ld. Howe tho' ready has not been able to sail, in short every accident seems to decide our ill fortune. I hope sincerely the events my not justify these remarks...

Another packet sails in three days. If we lose Rhode Island the only method we have of prosecuting the war must be from the [sea], and I now think that by employing your Land troops, to land, distress the coast with judgment, leaving it on the appearance of force, Washington's Army wou'd become of no use, then affairs might take a more favourable course.

2. Lieutenant Edward Fage, "Distance between the several works raised for the Defence of New-Port in Rhode Island."[3]

3 This document was in the now-lost papers of Charles Stuart with only a summary included in Anonymous, ed., *New Records of the American Revolution*, p. 57, titled "Some Distances in the Lines of Defence before the Town of Newport in Rhode Island, August 28, 1778." The two-page folio manuscript was endorsed "Distances taken by Lt. Fage, of the Artillery–with Prescott's compliments to Col. Stuart." The full transcription of Fage's reproduced letter here is from an unsigned and undated four-page copy of the letter in the Papers of Sir Henry Clinton, vol. 227:23, William L. Clements Library, Ann Arbor Michigan. Edward Fage was appointed a Cadet at the Royal Military Academy, Woolwich on 1 November 1755; He was commissioned as Lieutenant-Fireworker, 24 November 1760, 2nd Lieutenant on 9 December 1768; 1st Lieutenant on 17 June 1772; Captain-Lieutenant on 7 July 1779; Captain on 1 December 1782; Major on 5 December 1793; Lieutenant-Colonel on 9 September 794; brevet Colonel on 1 January 1800; Colonel, Royal Artillery on 9 April 1803; Major-General on 25 April 1808; Col-Commandant, Royal Artillery on 28 June 1808. He died at Greenwich, 2 September 1809. M.E.S. Laws notes in his book *Battery Records of the Royal Artillery, 1716-1859* (Woolwich, 1952), pp.61-65, that Fage was paid £39 3 6 for the loss of his baggage at Yorktown. It is not clear what the basis for this payment was. When Fage was promoted to captain in December 1782, he took command of a Royal Artillery Company at New York that had sailed with Admiral Graves's fleet from New York in October 1781 in the force intended to relieve Yorktown, however, they returned to New York when the fleet was unable to enter Chesapeake Bay.

Extent of the Line from the North Battery on the West side of the Island to the Redoubt over Easton's Beach on the South. .... 3110 yards

Measurement of all the Angles on the Banquette of the said Line 3885

From the North Battery to Eastons Redoubt in a strait Line 2500

From Eastons Redoubt to the 10 gun Battery at Green-end. The Right of the Exterior Line 2000

From Eastons Redoubt across the Beach to Walter Eastons house that is now burnt 1640

In the Exterior Line

From the 10 Gun Battery to the Seven	200
From the Seven to the Three	430
From the Three to the Center of Fort Fanning	600
From the Seven Gun Batt^{y} to Cards Redoubt	360
From the Three D^{o} … D^{o}	350
From Cards Redoubt to Dudleys	450
From Dudleys To Fort Fanning	380
From Dudleys to Bannisters Redoubt	360

Exterior Line

From Bannisters to Fort Fanning	210
From Bannisters to Irishes Redoubt	400
From Irishes to Fort Fanning	370
From Irishes to the foot of Little Tomini	750
From Irishes to the Center of Tomini	850
From the Center of Tomini to the wok at the foot of Little Tomini	200
From the Center of Tomini to the Center of Fort Fanning	800
From Fort Fanning to the Angle of the Barrier Redoubt in the Interior Line	1550

From the Angle of the Barrier Redoubt to the Center of Tomini Hill	1350

From the Angle of the Barrier Redoubt to the Center of Irishes Redoubt	1850
From the Center of Tomini directly West to the Water side	1100
From the Center of Tomini directly North to Coddington Cove	1440
From the Angle of the Barrier Redoubt to the Seven Gun Battery	1200
The Extent of the Swamp from the head of Eastons Pond to the Bridge in the East Road	1750
Saliant Angle of the Barrier Redoubt to the Town House at the Top of the Parade	1100
From the North Battery to Tomini Hill	1850
From Irishes Redoubt to the water side in Coddington Cove	1300
The whole extent with the Angles of the Exterior Line from the Ten Gun Battery to Tomini Hill	2527
From the 10 Gun Batty to Tomini Hill in the nearest Line	1700
From the Ten Gun Batty to Irishes Redoubt in a Strait line	1550
From the North Battery to Goat Island	1250
From Goat Island to Brentons Point	1200
From Brentons Point to the Dumplin ~~not~~ quite one mile	1700

Distance from our Exterior Line to the Rebel Batteries and Approaches on Honeymans Hill.

From our Seven Gun Battery to the Enemys Batt[y]

on their left on the Summit of Honeymans Hill .... 1200 yards

D°. to their left Batty advanced down the Hill ... 900

D°. Their right Batty D° 800

From Cards our Advanced Redoubt to their nearest Battery 700

From the Enemys Battery on the Left to that on their Right upon the Summit of the Hill 541

From their Right Batty to the one making by the East road of Seven Embrasures 1170

From that Battery by the East road to

Irishes Redoubt 1500

To Bannisters 1400

To Fort Fanning 1550

3. Lieutenant-Colonel Stuart to Lord Bute, New York, 4 September 1778[4]

4 Anonymous, ed., *New Records of the American Revolution*, pp. 57–58; also printed in The Hon. Mrs. E. Stuart-Wortley, ed. *A Prime Minister and His Son: From the Correspondence of the 3rd Earl of Bute and of Lt.-General The Hon. Sir Charles Stuart, K.B.* (London: Dutton, 1925), pp. 128–130

My Dearest Father,

I am this instant arrived here [in New York from Rhode Island], and understand that the Packet is to sail in four hours; you may easily conjecture how part of that time will be devoted, as it is some time since I last wrote.

I had arrived at Rhode Island on July 8th, the day after I had left Ld. Howe, and found Pigott's army in spirits. He had fortified the place by a line of Posts extending from Eaton's Beach on the east, round Banisters and Tomani Hills to the shore on the west of the Island, near Pest Island.

The Rebels had made their approaches on the left flank, from Honeymans Hill, but having chosen their ground injudiciously their nearest Battery was at 900 yards, with little probability of their being able to advance nearer on that side.

There was much cannonading on both sides, and during the night much Musketry owing to our Patrols, who, as we expected an attack, had orders to push near the Enemy's works. The Rebel Army consisted of 16,000 men; ours was scarcely 6,000 including sailors, who upon the first appearance of the French Fleet (July 29th) had had orders to burn 3 of our 32 Gun Frigates, 2 of 28 Guns, and 2 Ships of 18 Guns, besides Galleys—9 Ships in all, to prevent them falling into the enemy's hands.

As there was little probability of the Rebels remaining on the Island, I took a sketch of the works, acquainted Pigott with the reinforcement of 4,000 men ready to join him from the Sound, and returned; when I was surprised to find that the Genl. Was attending this small reinforcement &

we have just heard that the Enemy retreated upon perceiving this movement without loss.

A ship is also arrived from Lord Howe with accounts that he had taken a ship filled with pilots sent by the people of Boston to assist the French Fleet in entering that Harbour; some think from this circumstance that he had intercepted them before their arrival, buy by deserter's accounts they are already arrived.

You will see by the enclosed papers the continuation of the farce the Commissioners are playing; one has already march'd off the stage & I fancy it will not be long, from their last paper they all obtain a similar dismission.

They all call loudly for war, and have a thousand wise plans for the continuation of it. God knows how our affairs will be managed.

I have seen the orders from here which in my opinion ought not, at this juncture, to be obey'd. I hinted pretty clearly in a former letter the purport of them...but as the whole of the events of this War has and does depend upon the defeat or separation of Washington's Army; so ought we, if we have not the heads capable of undertaking the one, try to succeed by the other; and after securing our point for large vessels to the Northward, secure to the South & Bermudas, & by a plan which I intend to transmit to you by the next Packet, I think I can prove clearly that three fleets, well guarded by ships of War, containing three thousand men each, wou'd so harass the Continent by possessing one week, one province, one week another, that the Continental Army would dwindle to nothing . . .

The *Cork Victuallers*, which arrived some days ago, went by orders from home up the Delaware, and by the merest accident of taking a French brig off Reedy Island escaped being all taken; the *Lyoness* has, we fear, been told to take the same route, and will run the same danger.

4. Lieutenant-Colonel Stuart to Lord Bute, 16 September 1778.[5]

5 Anonymous, ed., *New Records of the American Revolution*, pp. 59–60; also printed in The Hon. Mrs. E. Stuart-Wortley, *A Prime Minister and His Son*, pp. 130–133.

Laurel Hill Camp, King's Bridge,[6] Sept. 6, 1778

My Dearest Father,

I believe in my last letter I acquainted you that General Sullivan had raised the siege of Newport Lines (Aug. 28th), and retreated from Rhode Island.

General Clinton is since arrived (Sept. 1st). From the particulars that I have been able to learn General Pigott, ill-informed, sent two columns on August 29th to attack the Enemy's rear before a man had retired, and instead of finding a few scattered troops, they fell in with a superior force under General Green, entrenched, and in all respects ready to receive them.

The consequence was that we were repulsed upon Quaker's Hill with the loss of nearly 400 men killed and wounded, and obliged to retreat to our Lines.

The Enemy retired next day without leaving any cannon, or other article that might give us the éclat of having drove them from the Island.

6 This position was in New York at the northern end of Manhattan near Spuyten Duyvil Creek. The Laurel Hill Camp was about 220 yards north of present-day Dyckman Street (West 200th Street) on the western side of modern Prescott Avenue.

The Corps that was intended for a reinforcement to Pigott went under General Grey[7] to Buzzard's Bay, up Dartmouth Creek to Bedford,[8] where they burnt a considerable number of Vessels, and destroyed a great quantity of stores. From thence, they went to Marha's Vineyard Island, where they at present are collecting stock for the Garrison of Rhode Island.

As packets sail so often, my dearest Father, it is a considerable relief, in our present melancholy situation, to be able to transmit frequently exact accounts of the very extraordinary events that happen.

Amongst the number that may be termed so, is Lord Howe's resignation in the present critical situation of our affairs.

---

7 Sir Charles Grey, later 1st Baron Gray in 1801 and Earl Grey and Viscount Howick in 1806. He was born in Howick, Northumberland, about 23 October 1729, the son of Sir Henry Grey, 1st Baronet. He was commissioned an Ensign in the 6th Regiment of Foot in England on 18 December 1746, a free death vacancy, and served at Gibraltar. He was promoted to Lieutenant in the 6th Foot on 23 December 1752, by purchase vice a retirement. He became Captain of his own new-raising 46th Company, of Marines on 21 March 1755, then transferred to be Captain of the 20th Foot in England on 31 May 1755, serving under Lieutenant-Colonel James Wolfe. Serving with Wolfe, he participated in the unsuccessful attack on Rochefort in 1757. He was promoted to Lieutenant-Colonel Commandant in England of his own new-raising 98th Regiment of Foot on 27 January 1761, a free augmentation vacancy. During the Seven Years' War he served as adjutant on the staff of Prince Ferdinand of Brunswick-Lüneberg, who commanded the Hanoverian Army of Observation. He was wounded at Minden in 1759, commanded a company at the battle of Kloster Kampen in 1760, was present at the capture of Belle-Isle in 1761 and at Havana in 1762, then served on the staff of Wilhelm, Graf zu Schaumburg-Lippe-Bückeburg, during the Spanish invasion of Portugal. He went onto the British Half Pay List when his regiment disbanded on 14 November 1763. He was promoted to Brevet Colonel on 20 December 1772 as ADC to HM King George III. From Half Pay, he became Colonel of the 28th Foot on 4 March 1777. During the War for American Independence, he was promoted to Major-General in America on 4 March 1777, then to Major-General in the British Army on 29 August 1777; he commanded the 3rd Brigade at the Battle of Brandywine in September 1777, at Germantown in October 1777, and Monmouth I June 1778. He was promoted to Lieutenant-General on 20 November 1782, then became Colonel of the 8th Light Dragoons on 13 July 1787; Colonel of the 7th Dragoon Guards on 17 March 1789. He was promoted to General on 26 September 1793 and commanded the West Indian Expedition to Martinique and Guadeloupe, He became Colonel of the 20th Light Dragoons on 4 November 1795. He served as Governor of Guernsey in the Channel Islands from 1797 to 1807. He was reappointed Colonel of the 8th Light Dragoons, 23 March 1797 and then Colonel of the 3rd Dragoons from 4 September 1799 until his death on 14 November 1807, after 47.6 years' service.

8 New Bedford, Massachusetts.

In all probability he had receive the consent to the request he made for leave to retire by the last Packet; he, however, kept it secret, sailed after d'Estaing, and by taking the track between Nantucket Shoals and George's Bank expected to be at Boston before him; unfortunately he was too late by a few hours, and only arrived off the Port to have the mortification of seeing an inferior French Fleet anchor at Nantucket.[9]

Not chusing to risqué an attack in that Harbour he returned, and resigned his Command. His Lordship's enemies are violent upon this occasion, and censure his whole conduct since the departure of his brother;[10] his friends even begin to wonder what private pique has been the cause of his delays.

I am ignorant of nautical matters, and therefore may say unwisely when I declare that I see no room for censure, except in the present circumstance. His Lordship knew that Gambier[11] was an idiot, and that Byron[12] was on his way to Halifax, and yet had not the Service sufficiently at heart to wait his arrival.

He knew that the French Fleet wou'd soon refit, and that our must remain inactive until Byron arrived—that during this time the French had the whole coast to range upon at pleasure, that Byron might be takem, Halifax attacked, and even this Port threatened; but nit to be tedious upon this subject, it would have been to his L—ps honor to have continued the nature of the war: to have considered the naval & military operations were too much blended to allow of that Department's being in the hands of one it was not design'd for & then in all human probability he

9 Nantasket Road in the outer harbor at Boston.

10 General Sir William Howe.

11 Rear Admiral James Gambier (1723–1789) was Commander in Chief of the North American Station from Lord Howe's departure in 1778 to 1779. He should not be confused with his nephew and namesake Admiral James, 1st Baron Gambier (1756–1833).

12 Vice-Admiral The Hon. John Byron (1723–1786), who was leading a squadron sent from England to support the squadron under Lord Howe.

wou'd have spared the General & and have prevented the hurt that this immediate change will occasion to the Service.

The Civil Commissioners[13] look upon a War of ravage or destruction to be the only means of effecting success in this Country; I doubt whether they have, on this pint, either power or understanding to advise: I fancy private resentment has drove them to adopt quite contrary opinions from those they came here with, not from a greater or more extensive knowledge of the Country, but from a deep sense of the insults they have received, and a determination to revenge them. We are all too apt to be influenced by our own inquiries when we decide on matters of Publick concern & their conduct has been so purile, that I cannot help giving this interpretation to their opinion.

As the mode of prosecuting the War is much the subject of our conversation & as conversation is likely to be the only employment of the campaign, I have made a very particular inquiry how far the people of this Country might be hurt by a war of Distruction; and find that with all our efforts they cou'd by no means be so materially hurt as people in general imagine.

There are only three towns of any consequence upon the Continent. N. York (Charleston was destroyed almost by fire), Philadelphia, and Boston are so situated, particularly the first, as to make it necessary to march with your whole army, and of course enter again into the same kind of War that has already been so unsuccessful.

There are some of the Colonies without any towns in our acception of the word…but if anyone would reflect on the extensiveness of the coast and the time it wou'd take to burn and destroy the breadth of ten miles from New Hampshire to Georgia, I believe they wou'd agree that this scheme could not be effectually executed.

---

13 The members of the Carlisle Peace Commission that had been sent to America to negotiate a peace with Congress. In an effort to influence public opinion they had warned of widespread destruction id the war were to continue.

We have no occasion to try whether acts of severity will cause these people to submit—wherever our armies have marched, wherever they have been encamped during this last campaign, every species of barbarity has been executed…We planted an irrecoverable hatred wherever we went, which neither time nor treasures will be able to eradicate. What, then, are we to expect from it, conciliation or submission?

There are some who argue that in case the Country is left, rigorous measure shou'd be adopted, in order to prevent too rapid growth of an Empire; there is no show of reason in this, but still the same argument of not being able to injure them sufficiently, is admissible; besides when you leave them Independent, the power they may arrive at, the consequences they masy be of, are to be considered…without rushing into measures that may ensure their determined Enmity for ever. In Political language we must weigh their Friendship and their Hatred & decide accordingly…

I call'd yesterday on Sir H. Clinton to learn his intentions, and found him perplexed beyond words; he told me he had received such orders since his taking the Command, and so contradictory to his first instructions, that he was determined not to remain.

He told me that Haldimand[14] made no requisitions for troops, and that we were stronger than ever on the Lakes.

14 Sir Frederick Haldiman (1718–1791) was a Swiss-born British Army officer. He served as Commander in Chief and Governor of the Province of Quebec (at that time including Ontario) from 1778 to 1786.

The expedition to the West Indies under Grant[15] is to start next week; our cousin,[16] goes with it; 5,000 British troops destined for this service. This is part of the orders from home...

A considerable part of the Army is to be station'd at Rhode Island under Lord C[ornwallis] for the winter,[17] the remainder here [in New York].

I now have hopes of being with you soon...Washington remains on White Plains.

15 Major-General James Grant, 4th of Ballindalloch (1720–1806).

16 William Medows (Eng.,b.'38): Ens, 50th Foot (in Eng.), 26.2.57, by means unk.; Lt, 50th Foot, 20.11.57, by purchase; to Capt, 4th Horse (in Ire.), 6.3.64, purchase; Maj, 4th Horse, 1.10.66, purchase; to Lt-Col, 5th Foot (in Ire.), 31.12.69, purchase; transfd to Lt-Col, 12th Lt. Dragoons (in Ire.), 17.9.73; transfd to Lt-Col, 55th Foot (in Ire, shortly to embark for America), 22.9.75; transfd to Lt-Col, 5th Foot (in Middle Colonies), 16.11.77; brevet Col, 25.11.78, as ADC to the King; to Col, 89th Foot (in the Leeward Islands), 15.11.80; Maj-Gen (local rank in East Indies), 17.1.81; Maj-Gen (in Army), 20.11.82; onto GB ½Pay List when 89th disbanded, 17.5.83; from ½ Pay as Col, 73rd Foot (in India), 11.8.86; served in India during 1782–92, where he was Gov of Bombay and C-in-C of the Bombay Army from Sept 1788 (transferring to the same commands at Madras in 1790). KB in Dec 1792; Lt-Gen, 12.10.93; full Gen, 1.1.96; Col, 7th Drag. Gds, 2.11.96 until dd at Bath, 17.11.1813.

17 As it turned out, Cornwallis was not inspired by neither garrison duty nor the progress of the war. At the same time, he had news that his wife was ill, so requested leave to return to England. He sailed on 28 November 1778 in the same ship with General Grey and the Peace Commissioners. Cornwallis did not return to America until 27 July 1779.

5. Lieutenant-Colonel Stuart to Lord Bute, 22 September 1778.[18]

18 Anonymous, ed., *New Records of the American Revolution*, pp. 60–61; also printed in The Hon. Mrs. E. Stuart-Wortley, *A Prime Minister and His Son*, pp. 133–135.

Camp Laurel Hill, New York, Sept. 22, 1778

I wrote you by the Packet a very long letter...I embrace this merely to recapitulate the substance of the last letter & to point out some errors that it contain'd.

Lord Howe it seems had possessed for some time a commission appointing Gambier Commander in Chief, but out of compliment to his Lordship there was no date, so that by dating the paper he had the power of resigning & making Gambier whenever he thought it convenient.

Four days after Lord Howe's resignation, Byron. With the *Culloden* and a frigate, was seen off the Hook,[19] but a severe gale of wind obliged him to take shelter in Rhode Island Harbour.

The arrival of Byron, instead of relieving us from our distresses as you naturally imagine, has rather involved us in more difficulties. He is arrived wit power merely to direct his Fleet in pursuit of d'Estaing, and Gambier, tho' a younger Admiral, remains in full power to command upon the American Station.

Added to this the French Fleet are un-attackable in their present situation at Boston, and Byron's Fleet are not only remarkably sickly, but have been so very ill-regulated as to have brought from England only six months' provisions and stores; in consequence of which (supposing them to meet with no accident) they will be obliged to return in the course of the winter.

19 Sandy Hook, at the entrance to New York Harbor.

You see, my dear Father, what infamous neglect Lord Sandwich[20] has been guilt of, and how little we are to expect true patriotism even from those who pretend to be guided by it.

I do not know how it will be the fashion to regard my Lord Howe upon his return, but in my opinion he has been the most guilty of men & if he had only the common feeling that a sense of his duty ought to occasion he wou'd be miserable at seeing the British Navy without a head to direct it & involved in the greatest confusion and distress; but if his private resentments have been the spring of this last action, I suppose a very different effect. We shall hear of his brother's wrongs and his own imaginary injuries, & the present melancholy situation of the fleet, instead of appearing against him will be quoted by himself as a proof of the injustice done to the service in permitting to return.

My rage has already made me say too much in my last letter upon this subject, but I cannot help it. I wish to God I was out of this Country, not to be eye-witness to these scenes. When I see them, I cannot help feeling & the only resource I have is unbosoming myself to you.

Grey is returned without having met with opposition; he brought from Martha's Vineyard Island 7,000 sheep, 4 or 5 of his transports were separated from in a gale of wind, we are much alarmed concerning them.

I am afraid that while we remain in this inactive situation the French ships, aided by the New Englanders, will invest Halifax; however, McLean[21] is arrived with the reinforcements, and if he has taken the precaution to fortify some islands that command the entrance of the Harbour, I should think that Byron wou'd be able to sail to its

20 John Montagu, 4th Earl of Sandwich (1718–1792) was First Lord of the Admiralty from 1771 to 1782. This was his third time that he held the office.

21 Lieutenant-Colonel Francis McLean (ca. 1717–1781), commanding the 82nd Foot, arrived in Nova Scotia in September 1778. General Clinton gave him the local rank of Brigadier when he replaced Major-General Eyre Massey as military commander at Halifax.

relief before they can effect the reduction. We know that the French will find it difficult to remain in Nantucket [Nantasket] the winter, and that if they are unable to take Halifax they will in all probability return to France.

We have various, but, I believe, very exaggerated, reports of disagreements between the French and Americans. From what I can collect from them, I find that the Americans have not approved of their conduct, and the French despise them most heartily, but Congress, knowing the importance of the Alliance, do every condescending act to strengthen it.

The Rebels, if they dared to utter their sentiments, have very forceable reason against their Friendship, they were taught to expect relief where they have found additional distress, and the following ridiculous anecdote shows more completely what a burthern they (the French are) to their allies.

When the French first arrived off the Delaware, Congress employ'd all their friends as contractors to supply the Fleet, imagining thereby to enrich their favorites and bring a quantity of specie into the country. D'Estaing received every article that was collected for them, but when their exorbitant accounts were deliver'd for payment, instead of disbursing as they expected, he gave them a receipt and an order upon Congress for the Whole!—which not withstanding theor reluctance they were obliged to accept.

General Grant with 5,000 British sails the first week in Oct. for the west Indies; my regiment was to have gone on that service, but a Friend of mine found means to have another appointed without my knowledge; notwithstanding that I detest the man that is to command and dread the climate, I long to go there, and I own I was a little hurt at this act of friendship.

Johnstone[22] and General Robertson[23] are going home; the first has not only acted a strange part in Publick, but now he has resigned his Commission makes no scruple of trying to prevent any effect from the Commission by declaring that Lord H[owe] not only means, but has all

---

22 Captain George Johnstone, RN, (1737–1787), who was a member of Parliament and serving as a member of the Carlisle Peace Commission to America. Johnstone had sailed in merchant ships as a boy, then joined the Navy and served with distinction in the Seven Years War. He was commissioned a lieutenant on 9 October 1755, promoted to commander 9 February 1760, then Captain, 11 August 1762. Placed on half pay at the end of the war, Lord Bute recommended him as Governor of the new British colony of West Florida, where he served from 1763 to 1767. His plan to start a war with the Creek Indians ran counter to government policy and the ministry recalled him. Through Stuart's brother-in-law, Sir James Lowther, Johnstone then unsuccessfully stood for Parliament but eventually took up Lowther's pocket borough of Cockermouth. He opposed Lord North's American policy, but also opposed the Rockingham's support for American independence. He was ridiculed for criticizing the appointment of Carlisle to head the peace commission, then joining it. Hoping to negotiate a settlement with the Americans, he tried, but failed, to supplement the formal negotiations with his own private talks. With this failure, he resigned from the commission and returned home. He blamed Lord Howe for the failure of naval operations in America and through Germain and Sandwich should resign. Unable to bring any opposition members to support the government, he received no political office, but returned to sea command, where he was eventually given command of an expedition to the Cape of Good Hope in 1781.

23 James Robertson was born in Scotland in 1717. Commissioned a 2nd Lieutenant in the new-raising 2nd Marines, 29 November 1739, by raising men for rank. He was promoted to 1st Lt, 2nd Marines on 19 April 1741, a free death-vacancy. Promoted to Captain, 2nd Marines, 12 September 1745, by purchase, he went into British Half Pay when his regiment disbanded, 9 November 1748. He exchanged from Half Pay as Captain in the 30th Foot, 15 April 1749. He was promoted to Major in the 38th Foot, 23 December 1755, then became Major in the new-raising 2nd Battalion, 62nd/60th Foot, 26 December 1755. Receiving a brevet promotion to Lieut-Colonel, 8 July 58, he served as Deputy Quartermaster General in America after Loudoun appointed him to that post on 8 March 1757 and served in that position until 1763. Thereafter, he served as Barrack-Master General in America from 1765 to 1771. He was promoted to Lieut-Colonel in the 55th Foot, 29 October 1759, a free death-vacancy, then he transferred to Lieut-Colonel in the 15th Foot, 25 February 1760, exchanged to Lieut-Colonel, 16th Foot, 17 August 1768. He became a brevet Colonel, 25 May 1772, in a general brevet promotion. He was promoted to Maj-General (in America), 1 January 1776 and to Lieut-Colonel Commandant in the 2 battalion, 60th Foot, 11 January 1776, a free dd-vacancy. In 1778, he was promoted to Colonel-Commandant. He became Major-General in the Army, 29 August 1777 and Colonel, 16th Foot, 14 May 1778. H was promoted to Lieut-General, 20 November 1782; and 4 March 1788 in Wimpole Street, Cavendish Square, London.

along meant to give up this Country; that some very infamous papers said to be his are Ld. N[or]th's.

I am sorry to inform you that Robertson has by no means the understanding he had & Johnstone, availing himself of the decline of this poor man & the abilities he was once thought to have, has entered into a strict intimacy with him, & I make no doubt, in assertions, will make use of his name and appeal to his evidence. I have cautioned Robertson to beware of him, but as I know his weakness & the other's art, I despair of his taking my advice & only beg that you will be cautious in believing what he says, or in relating anything you might not wish to have repeated.

Washington has moved from Whiteplains to Peeks Kill on account of forage; this makes no alteration in our situation; Lord Cornwallis is destined with part of the Army for Rhode Island, and if the same inactivity continues, I hope to sail for England the end of next month.

Addendum

The notes below accompany notes of the same number on pages 43, 65, and 95 respectively.

16 Henry Barry was born in England in 1750. The son of Captain Robert Barry and his wife Sarah, of Worcester, Henry Barry was commissioned as 2nd Lieutenant in the 85th Regiment of Foot in Portugal on 22 February 1763, a free vacancy. He went onto the British Half Pay List when the regiment disbanded on 24 May 1763. On 11 March 1768. he transferred to the 52nd Foot as an Ensign at Quebec. There, he was promoted to Lieutenant in the 52nd Foot on 23 September 1772, by purchase. Barry was commissioned in a provincial captaincy of 26 December 1775 in the new-raising Loyal Nova Scotia Volunteers but resigned from it shortly before 12 May 1776. In this period, he published two anonymous pamphlets. After his regimental promotion to Captain-Lieutenant in the 52nd Foot on 4 January 1777, by purchase, and his Army promotion to Captain on 8 May 1777, the 52nd Foot returned to Britain for recruiting duty in late 1778. Barry returned to America in 1780 as deputy to Lord Rawdon, the Adjutant-General of British Forces in America. At the Battle of Eutaw Springs in September 1781, American forces captured Barry and held him as a prisoner of war until March 1782. Following a brief return to England, he became a Brevet Major in the Army on 19 February 1783. He remained in India for some time after the peace but had returned to England by 1787. He was promoted to Lieutenant-Colonel in the 39th Foot in Ireland on 8 December 1790. He resigned on 31 March 1793 with 25.3 years of service, but was then promoted to Colonel on 19 July 1793, then joined Major-General Francis Rawdon-Hastings, Lord Moira, on his expedition to Ostend in June-July 1794. He died at Bath on 2 November 1822.

20 Lieutenant John Knowles, Royal Navy, served as the Agent for Transports at Newport, Rhode Island, from December 1776 to August 1778. Knowles was commissioned Lieutenant on 19 January 1759. He was promoted to Commander on 29 September 1778 and relieved Commander James Watt in command of the 8-gun fireship Sulphur, remaining in her only until 13 January 1779. Promoted to Captain on 1 July 1780, he commanded the 24-gun frigate Fowey for a year until 22 June 1781, then the 50-gun Dutch prize Rotterdam, and the 70-gun Spanish Prize Diligent as a guardship at Portsmouth until April 1783. Not employed at sea again, he was promoted in retirement to Rear-Admiral on 14 February 1799. Of the 20 officers who had been agents for transports during the American War, he was one of only two men who were eventually promoted to flag rank.

31 Major-General Sir Robert Pigot, second Baronet, was born on 20 September 1720. General Pigot was the brother of Sir George Pigot and Admiral Hugh Pigot. Robert Pigot was commissioned an Ensign in the 31st Regiment of Foot in 1741 and served with his regiment in Flanders. He was promoted to Lieutenant in 1744 and was present at the Battle of Fontenoy on 11 May 1745. In September 1745, he returned to England, then in 1749 served in Minorca and then,

Scotland. He was promoted to captain on 31 October 1751. In 1758, the 70th Regiment of Foot absorbed his battalion of the 31st Foot and on 5 May 1758, Pigot was promoted to Major. He joined the 38th Foot and advanced to Lieutenant-Colonel on 1 October 1764. In 1768, Robert succeeded his elder brother George as Member of Parliament for the venal and expensive constituency of Wallingford in Berkshire. He held the position until January 1772, when he was appointed Warden of the Mint. His attempt to stand for re-election to Parliament failed in 1772 and, at Stafford, in 1774. Promoted to Colonel in 1772, he and his regiment were ordered in 1774 from Ireland to North America, where the regiment was posted to Boston to quell the local disturbances. On 19 April 1775, he joined the relief force under Brigadier Hugh, Earl Percy, to support Lieutenant-Colonel Francis Smith's withdrawal from Lexington and Concord. Pigot distinguished himself at the Battle of Bunker Hill in June 1775, when he commanded the army's left wing with the local rank of Brigadier-General in the attack on the American redoubt on Charleston Heights. In a series of assaults, Pigot broke American resistance at great cost to his own men. To honor his bravery and distinguished service, King George III promoted him to colonel of the 38th Foot. He went with his regiment to Halifax, when the British Army evacuated Boston, then went to New York, where he commanded the second brigade at the battle of Long Island on 27 August 1776. While still in America on 11 May 1777, Pigot succeeded on the death of his elder brother, George, to his baronetcy, inheriting at the same time his brother's estate at Patshull, Staffordshire, and a third of the £30,000 Pigot diamond. On 15 July 1777, Pigot took command of the British garrison at Newport and on 29 August 1777 was promoted to Major-General. In March 1778, he was under consideration to be named Howe's second in command, but this did not materialize. While in Rhode Island, the bon vivant Pigot was generally unhappy as he missed the social life in New York. Pigot relinquished his command in October 1779 and returned to England but saw no further active service. He was promoted to Lieutenant-General on 20 November 1782. After a long illness, he died at Patshull on 2 August 1796.

Index

The index references to the diary are confined to the transcription and its notes. Where the original manuscripts uses alternative spellings for modern placenames the form used in the manuscript is included in parentheses.

C

D

E

F

G

H

I

J

K

L

M

N

T

U

V

W

Y

About the Editor

Professor John B. Hattendorf is the Historian of the Rhode Island Society Sons of the Revolution. In addition, he also serves as Historian of the Rhode Island Society of Colonial Wars and Rhode Island Society of the Cincinnati. He is the Ernest J. King Professor Emeritus of Maritime History at the U.S. Naval War College's eponymous John B. Hattendorf Center for Maritime Historical Research. While holding the E. J. King chair from 1984 to 2016, he was chairman of the Naval War College's Advanced Research Department, 1986-2003; chairman, Maritime History Department, and director of the Naval War College Museum, 2003-2016. As a naval officer, 1964-1973, he saw combat action while serving in destroyers during the Vietnam War. He holds degrees in history from Kenyon College (A.B., 1964), Brown University (A.M., 1971), and the University of Oxford (D.Phil., 1979; D.Litt., 2016). His numerous awards include the Anderson Medal for Lifetime Achievement from the United Kingdom's Society of Nautical Research (2017), the U.S. Navy's Distinguished Civilian Service Medal (2016) and Superior Civilian Service Medal (2006, 2016), the American Library Association's Dartmouth Medal (2007), and the Caird Medal of the National Maritime Museum, Greenwich (2000). Additionally, in 2011 the U.S. Naval War College named its Prize for Distinguished Historical Research in his honor and in 2017 its Center for Maritime Historical Research. He was inducted into the Rhode Island Heritage Hall of Fame in 2019. He is the author, co-author, editor, or co-editor of more than fifty volumes, including the four-volume *Oxford Encyclopedia of Maritime History* (2007).